STANDAR

To Mónica del Pilar

Peter Bunyard

The Breakdown of Climate

Human Choices or Global Disaster?

Floris Books

First published in 1999 by Floris Books

British Library CIP Data available

ISBN 0-86315-296-1

Printed in Great Britain
by Page Brothers, Norwich.

Contents

Acknowledgments

This book has taken several years to come to fruition and I would like to thank Teddy Goldsmith for supporting and encouraging me during this time. For more than thirty years, he has inspired me with his vision and indomitable determination to leave a safer world for those to come. I also thank all who have worked so selflessly for the *Ecologist,* ensuring its survival during difficult times. In particular, Simon Retallack, who proved to be an inspired editor and writer during the difficult compilation of the special 1999 issue of the *Ecologist*, on the climate crisis. Stephanie Roth also provided extraordinary support in keeping us all abreast of current events.

I owe a great debt of gratitude to Jim Lovelock, whose extraordinary concept of Gaia has transformed the way in which we view life on this planet. Also, I would like to thank the newly established Gaia Society, with due credit to Philip George, Tom Wakeford and others at the University of East London, who formed the society. Through them I have met an extraordinary group of scientists, activists and philosophers.

I am particularly grateful to Brian Goodwin and Mae-Wan Ho who, together with Peter Saunders, impressed upon me how much more there is to life than a package of 'selfish' genes and globules of cytoplasm.

No evolutionary theory would be complete without recognizing Lynn Margulis, who, more than anyone, has made us realize our humble origins from bacteria.

I would like to thank Stephen Harding of Schumacher College for his friendship and open-mindedness in discussing evolution and Gaia, as well as enjoying many musical hours together. My appreciation extends to Paco Peña, who, while always maintaining an interest in what I was doing, introduced me to a different, but just as complex and mysterious, world, through his wonderful music.

To my family and friends, my gratitude for their support and advice, particularly Jeremy Faull, Director of The Ecological Foundation.

Others who have been generous with their time and insights include Richard Betts at the Hadley Centre, Mike Fasham and his colleagues at the Oceanography Centre in Southampton, and not least Michael Whitfied, Director of the Marine Biological Association at Plymouth.

To all the others not mentioned here, who have contributed in innumerable ways to the writing of this book, my grateful thanks.

Introduction

The Modern World

Today we live increasingly in a man-made world that we isolate as far as possible from the elements. Double, even triple, glazing on our windows, plus a host of contraptions such as boilers, heat pumps and air-conditioners, all contrive to keep us at comfortable temperatures and humidities, whether we are at work, at home, or travelling in cars, trains, or aircraft. As long as everything works and the engines and power supply keep running, we could be living in any climate under the Sun. Ironically, our attempts to control our immediate environment have exacerbated the very problem we are trying to avoid — global warming.

Yet, however much we try to generate our own climate, the reality is that our very survival depends on our relationship with the natural climate. What we can grow in our fields, how and when we harvest, what water resources we have, whether we are likely to suffer drought or flood, how cold or warm the winters are, these issues are all intimately linked to climate. Ever since humans began to migrate, leaving the warm, equatorial belt of Africa for more extreme climes, they have become more exposed to the vagaries of weather. In addition, the world population is increasing rapidly, especially since World War II, and a growing population needs a reliable food base to sustain it, which requires climate to remain within reasonable limits for animal and crop husbandry.

The Past

We know from history that the idea of a stable, unchanging climate is illusory. The climate has changed dramatically even over the past thousand years. If we had had access to instantaneous news from every corner of the planet when we were roaming the Earth as hunter-gatherers, we would have known that somewhere, at that moment in time, people were suffering from some unprecedented climatic disaster.

Once the Neolithic revolution was underway with the domestication of crops and livestock, man believed in the potential consistency of climate, in order to grow crops. When the climate went seriously wrong man thought that perhaps he had broken cosmological laws and disrupted the natural order. Sacrifice, atonement and soul-searching became man's way of restoring the balance and getting the climate back into kilter.

The Present: 1990–95

Worldwide, the past decade has witnessed the warmest average temperatures for a century. 1990 broke one record after another with summer temperatures in England rising to levels usually associated with the Mediterranean; Cheltenham, in Gloucestershire, recording a high of 37.1°C. In France, the hot dry summer left 3,000 km of rivers virtually empty and triggered widespread forest fires in the southern part of the country, as well as in Italy, Spain and Portugal. 1991 might have followed suit if it had not been for the massive eruption of Mount Pinatubo, in the Philippines, which created aerosols high in the atmosphere that reflected sunlight back into space, and so cooled the Earth. In 1993, the Mississippi River burst its banks and inundated vast areas of land following the worst rains for 500 years.

By 1994, the aerosols from Mount Pinatubo had been washed out of the atmosphere and global temperatures again began to rise. In central India, that summer, a heatwave left many people and livestock dead, as temperatures stayed at 46°C for several weeks. Japan had the hottest summer on record, having to send tankers to Alaska for freshwater. In England, scientists at London's Kew Gardens pronounced that the summers had definitely become *Mediterranean*, with farmers and gardeners having to cope with pests and diseases more typical of southern France.

In northern Europe in 1995, the hottest summer since records began in the early eighteenth century, was contrasted by some of the most extensive winter floods for a hundred years; in January, the people of the Netherlands battled day and night to reinforce the dykes, while in France, the worst floods of the century destroyed more than 40,000 homes. In August, scorched brown pastures in England's green and pleasant land, meant that livestock were fed hay and silage that should have provided for the winter months, and people queued at standpipes for their ration of water. By the end of

1995, meteorologists at the UK Met Office, claimed it had been the warmest year in Britain since records began in the late seventeenth century, despite a severe chill over Christmas.

The *Independent on Sunday* also portrayed 1995 as 'The Year of the Hurricane,' pointing out that by early October sixteen tropical storms, all with wind speeds above 120 kph — and therefore designated as hurricanes — had arisen in the Atlantic, leaving a swathe of destruction in their wake. Hurricane Opal attained wind speeds approaching 240 kph, killing seventeen people and leaving damage amounting to nearly $2 billion in the United States alone.

However, 1997 and 1998 will be known as the years in which lingering doubts about global warming and its impact on climate were really brought home by an even more dramatic torrent of disasters (see Chapter 4).

El Niño

Why did these climatic changes occur? Why did routine annual burnings go so horribly wrong? The reason is linked to a climatic phenomenon that periodically affects the ocean currents of the Pacific. Most years the trade winds blow across the Pacific Ocean from east to west, driving the surface waters with them, which end up in the seas of South-East Asia creating the monsoons. The rains dampen the rainforests, thus preventing living trees catching alight by fires that have been lit in the clearings. But in 1997, the trade winds weakened and failed. The surface waters of the tropical Pacific changed direction and began collecting, along the flank of South America. The high pressure zone — bright weather and clear skies — now centred itself over South-East Asia, while the low pressure zone — usually associated with monsoon rains — shifted to the deserts along the coast of Peru, which is where the rain fell, thousands of miles away from its usual destination.

An obvious reaction to these weather changes would be to believe that the deforestation of the past few decades caused the failure of the summer rains. Yet, as Peruvian fishermen know, the switch in the Pacific currents, and the strengthening or weakening of the trade winds, is a natural phenomenon which takes place every few years. For centuries, the fishermen have depended on anchovies for their livelihood, which normally thrive in cool nutrient-rich waters brought by the Humboldt current that flows along the coast from Antarctica. When the Pacific currents change direction, the Humboldt current is held at bay, replaced instead by the warm, nutrient-poor tropical waters that sweep in from the western Pacific. The anchovies dive in search of food and oxygen, and as far as the fishermen of Peru are concerned, their catch virtually vanishes.

The fishermen dubbed the warm water current *El Niño* — the Christ Child — because it manifested itself in the days around Christmas. The name has stuck, immortalized by Sir Gilbert Walker, the English meteorologist, who first studied the phenomenon seventy years ago.

Climate Change

Although El Niño used to appear every few years, at most twice every decade, lasting for just one year, the phenomenon has become more common. Over the past two decades it has even appeared several times in a row. Its effects have also become more pronounced, for example, 1997 and 1998, when countries across the planet experienced drought when normally they would have rain and vice versa. Colombia, for instance, suffered a devastating loss of crops due to drought, whereas farmers in northern India experienced equally devastating losses from unprecedented rain that destroyed harvests and prevented sowing.

Are these changes in the strength and frequency of El Niño simply a consequence of natural variations in the Earth's climate? Or are they a reflection of what man is doing to the planet?

Our activities are clearly having a great impact on the environment. We are clearing forests, draining marshes, paving over land, creating megalopolises, generating superhighways for a growing torrent of traffic, and variations in climate are definitely affecting agriculture. The question is whether our activities are now unsettling the climate to the point where we can no longer be sure of weather and seasons.

Sudden switches in climate have always been potentially disastrous for human populations; northern Europe suffered devastating famine in the late Middle Ages when the warm climate of the previous centuries vanished abruptly, and people in Scotland and Scandinavia found themselves with nothing to eat but the bark of trees. But never before have we been so close to the limits in terms of the amount of land used and the availability of fundamental resources such as water.

The Californian biologist Paul Ehrlich points out that man, with his crops and livestock, has now taken over the management of forty per cent of the total land surface's capacity for production. Major catastrophes aside, the world's population is expected to double by the middle of the next century, and unless we double food production from currently available land it will mean that we will have to manage eighty per cent of the total land surface, involving intensive production in marginal places such as high mountains, and swamps. Climate change could put paid to any plans we have for finding the resources to feed the population in the future.

Yes, the climate is changing; that is the official news coming from the scientists who the United Nations first brought together in 1988 as part of the Intergovernmental Panel on Climate Change — the IPCC. In 1990, these same scientists reported that more than a century of industrial activities could be responsible for the 0.5C average rise in the surface temperature of the planet. In particular they blame emissions of carbon dioxide from industry as a prime cause of the warming of the Earth's surface. If their ideas are half right, we have only just begun to experience the warming. It could become manifestly worse.

As the *Sunday Telegraph* (7 February 1999) points out, 1998 was the worst to date for climatic disasters, with worldwide estimates of damage from natural disasters amounting to more than $50 billion a fifty per cent increase from 1997. Paul Kovacs, an analyst with the Insurance Board of Canada, remarked, 'Every five years or so the costs of weather-related disasters have doubled, with the last three years the costliest in history for the Canadian insurance industry.' The Munich Reinsurance Corporation — one of the world's largest underwriters — estimates that over the past ten years the cost of all natural catastrophes, many relating to climate, has risen 85 times above the cost for the 1960s, adjusted to present values.

Causes of Climate Change

The Earth's surface is kept warm because of greenhouse gases, so-called because like a greenhouse, which lets light through but retains heat. As a result, Earth's surface temperature is considerably higher than if no such gases were present. From physics we know that without greenhouse gases in the atmosphere the average surface temperature of our planet would be -18°C. The temperate zones would be uninhabitable: even the Tropics would be freezing.

Each year, since the mid-nineteenth century, we have been burning billions of tonnes of fossil fuels, as well as destroying forests through slash-and-burn techniques. As the carbon in the fuel is burnt carbon dioxide is released, which is a greenhouse gas like other atmospheric gases such as methane and most importantly, water vapour. That means we are contributing to the levels of natural greenhouse gases, as well as adding some of our own, such as CFCs (chlorofluorocarbons). The total amounts of carbon dioxide that we have added to the atmosphere are staggering: 190 billion tonnes over the past fifty years. In that period we have emitted three times the carbon dioxide released by mankind over rest of time put together. Since World War II alone we have increased the level of carbon dioxide by nearly one quarter. Not all that carbon dioxide, has stayed there. Approximately half appears to have left the atmosphere, absorbed into the oceans, soils and perhaps even buried in sediments. We are lucky that so much has gone from the atmosphere but may be wrong to assume the natural sink will always operate effectively to limit the damage we are imposing. How can we be so sure that the recent past will reflect the future?

In 1995, after years of intense investigation of the atmosphere, the oceans, the continents and the polar regions, scientists assigned to the IPCC confirmed that not only had we caused current global warming, but that if we continued to pour our waste gases into the atmosphere, we could expect temperatures to soar over the coming century. No one can predict absolutely what might happen, but we could experience dramatic changes to local climate, with severe consequences for property, lifestyle and essential food production, let alone for wildlife.

The correlations between warming and climate catastrophe are clear. In north-west Canada and Alaska mean annual temperatures a metre below the soil surface have risen a full degree since 1989. There, as in Siberia, this is causing the permafrost to melt, releasing its store of methane, which is adding to the ever-growing increment of greenhouse gases. We are now seeing plants, insects, birds and mammals, including pests and agents of disease, migrating northwards into regions previously too cold for them. In low-lying islands in the South Pacific farmers are having to abandon their fields because of sea level rise and some islands have had to be vacated.

According to the IPCC's latest coupled ocean-atmosphere models, by 2080, with double pre-industrial CO_2 concentrations, we could see an average global temperature increase of 2.5°C, with perhaps 4°C over land masses, particularly in the northern high latitudes, 3° to 4°C over parts of the Arctic or Antarctic, and considerable regional varia-

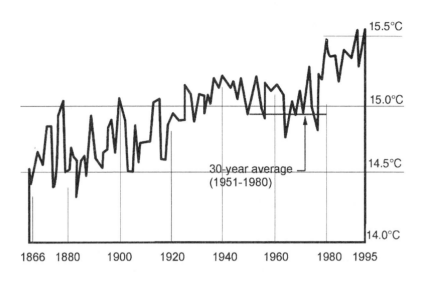

Figure 1. Annual average global surface air temperatures. (Source: NASA Goddard Institute for Space Studies).

tions. What is more, a second doubling of pre-industrial levels of CO_2, which could occur by the end of the 21st Century, could lead to a catastrophic rise of as much as 10°C.

The Case against Global Warming

Despite the growing scientific evidence, a small number of self-created experts, who represent the powerful interests of the fossil fuel lobby in the United States and the oil-rich nations of the Middle East, continue to lecture the world that current global warming either has nothing whatsoever to do with human activities, or is simply not happening. Sadly, their influence over governments, especially over the Republican majority in the US Congress, is preventing vital, immediate action to stem the ever-growing atmospheric carbon emissions.

Patrick Michaels, one of the most vociferous climate change sceptics, spokesman for the Global Climate Coalition and Professor of Climatology at the University of Virginia, has travelled the world dismissing the idea of global warming. However, even he has been put on the defensive as science becomes more sophisticated and the models show vastly improved correlations. A decade ago, General Circulation Models (GCMs), as they were then, indicated that the global average surface temperature should have risen by one degree centigrade over the past century. The global warming sceptics were quick to point out that by the climatologists' own admission the actual overall temperature increase was little more than half a degree, giving the sceptics considerable ammunition for deriding the efforts of IPCC's scientific committee. To add fuel to their cause, satellite measurements revealed that the lower atmosphere, about 3.5 kilometres up in the troposphere, had cooled by 0.05°C. 'The theories were flawed,' pronounced the sceptics. Instead of warming, the Earth was cooling. By drawing attention to the apparent cooling some kilometres up they ignored the obvious discrepancy that average surface temperatures were increasing by 0.13°C per decade, and the lower stratosphere was cooling by as much as 0.5°C per decade: both facts evidence of significant surface warming.

This discrepancy has now been resolved. Frank Wentz of Remote Sensing Systems in Santa Rosa, California, points out that the data being beamed down from satellites had been interpreted as if the satellites were in a stationary, unchanging orbit. No one, he remarks, took account of the slippage over time of the satellites as they were inexorably pulled in closer to the Earth because of atmospheric friction. The slippage of 1.2 kilometres every year gradually alters the angle at which the measurements are made, therefore giving a spurious result. Wentz re-did the calculations to account for the real angle and discovered a warming trend of 0.07°C per decade — just what would be expected from the readings from other strata in the atmosphere.

To add weight to this evidence, scientists from the British Antarctic Survey have discovered that the outer atmosphere is shrinking at the rate of one kilometre every five years, because, with more heat trapped at the surface, less is reaching the outer atmosphere, which is, in fact, getting colder.

Meanwhile, climatologists at the UK Met Office's Hadley Centre have explained the discrepancy between the 1°C warming shown by the earlier models and the actual 0.6°C actually warming experienced across the globe. Sulphur dioxide is also emitted when fossil fuels are burnt and it acts as an *aerosol* which reflects light back into space. Its cooling effect therefore tempers the warming that would result from the greenhouse gases alone. Furthermore, the sulphur

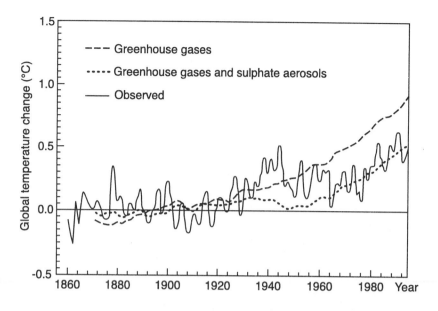

Figure 2. Simulated global annual mean warming from 1860 to 1990, allowing for increases in greenhouse gases only, and grennhouse gases and sulphate aerosols, compared with observed changes over the same period. (Source: Houghton et al (Eds), Climate Change 1995, *Cambridge University Press).*

dioxide has a short lifetime in the atmosphere amounting to several weeks instead of the decades or even centuries, associated with the greenhouse gases. In times of past economic recession, like the Great Depression, less fossil fuels are burnt and less sulphur dioxide is released. The cooling effect is therefore relatively weaker than the warming brought by past releases of greenhouse gases and, as the Hadley Centre climatologists discovered, under such circumstances, the general climate tends to show a warming spell.

The Future

Viable agriculture needs a stable climate. If we cannot anticipate from one year to the next what and when to sow and what sort of harvest to expect because the climate is going through unpredictable convulsions, then we are in serious trouble. According to current general circulation models, the worst impact on agriculture will be in Africa, the Middle East and the Indian subcontinent. We are told that most of Europe and the humid tropical countries of South-East Asia, will benefit from global warming, at least until the 2080s (see chapter eleven). With trade liberalization gathering steam, the premise is that more food will be made available — for those who can afford to shop. Until we have a better grip on what contributes to the world's climate, we are left with little alternative but to take the IPCC projections at face value and hope that life will continue to function effectively as a vital, integral part of the carbon cycle. Past as well as present greenhouse gas emissions will affect global warming in the future, because of the time lapse required for the gases to disappear from the atmosphere. Between forty and sixty per cent of the carbon dioxide currently released into the atmosphere, is expected to take 30 years to be removed. Consequently, the IPCC has taken into account different scenarios of future energy use and just how that energy is produced. The IPCC assumes that growth in energy demand will continue as industrialization and the market economy spreads throughout the world, especially in developing countries and the Third World. Just how that growth in demand will be met, whether from fossil fuel burning, renewable energy sources such as wind, hydro and solar power, which turn sunlight directly into electricity, or from an expanded nuclear power base, will clearly affect global emissions of fossil fuels.

In the meantime we are still pumping out greenhouse gases. Using 1990 as the baseline, the agreement thrashed out at the Kyoto meeting of the IPCC in December 1997, called for industrialized countries to cut back on greenhouse gas emissions by an overall 5.2 per cent between the years 2008 and 2012. Countries such as Britain will achieve substantially greater cuts, particularly by switching from coal to natural gas for power generation, since natural gas has a higher calorific content than coal. With only four per cent of the world's population, the United States is responsible for nearly one quarter of the world's greenhouse gas emissions, and its citizens emit five times more greenhouse gases on average than anyone else in the world. In 1992, the US administration voluntarily agreed that by 2000 it would cut its greenhouse gas emissions to 1990 levels. Today, those emissions stand at thirteen per cent higher than 1990 levels and in a decade, based on current projections, are likely to be thirty per cent higher.

Even if industrialized countries achieve reductions over the next 10 years, developing countries are rapidly increasing their greenhouse gas emissions. China, for instance, is building a number of new coal-fired power stations, and is currently the world's number one coal burner. Currently, greenhouse gas emissions are increasing at rates even faster than the more extreme of the IPCC's business-as-usual scenarios (see Appendix I). Unless drastic action is taken now carbon dioxide levels in the atmosphere will double every 27 years. That is a very different scenario from a doubling of carbon dioxide by 2080, on which most policy makers are basing their understanding of what the future holds in store.

The Consequences

With current trends, instead of the maximum 4.5°C average rise in temperature by 2100, indicated by the IPCC for a business-as-usual projection, average temperatures could rise by 10° or even 14°C. The consequences in terms of climatic disruption would be enormous. Those most at risk from global warming are countries with considerable areas of low-lying land, such as many islands in the Pacific, Caribbean and Indian Oceans, or regions with large delta areas, such as Bangladesh (see Chapter 6). In general, such regions have relatively low per capita incomes and a long vulnerable coastline in relation to land mass. Since the United Nations conference in Rio de Janeiro 36 of these independent countries formed the Alliance of Small Island States (AOSIS), to encourage industrialized countries to commit themselves at least to a twenty per cent reduction in greenhouse gas emissions by the year 2005, using 1990 as the baseline.

Global warming has its paradoxical side, perhaps the most bizarre being the likelihood that the Gulf Stream will stall as an excess of freshwater flowing into the North Atlantic and an increase in melting sea ice cause the density of surface waters to fall, so that they no longer sink (see Chapter 6).

Gaia

In the late 1960s the British scientist, James Lovelock, came up with a thesis, which he later called Gaia, purporting that life and the atmosphere were bound together in a self-regulating system. Scientists have increasingly discovered the influence of life in many of Earth processes, including the formation of continents and possibly tectonic movements of the crustal plates. The Gaia thesis suggests that life is not a mere passenger on this planet but is indissolubly connected to the processes of transformation that take place all over the Earth's surface. It is becoming increasingly apparent that life itself maintains suitable conditions on the planet for life. Thus, a good climate is a necessary condition for life, and conversely, a planet that is replete with healthy living systems has an equitable and well-regulated climate.

It is as yet unknown how life will react to changing temperatures and climate change, and yet life itself is crucial in affecting the balance between greenhouse gas emissions and their removal from the atmosphere. The assumption is that living systems will continue to draw carbon dioxide from the atmosphere and will compensate to some extent for man's profligacy in exploiting natural resources. It is perhaps on this issue that the IPCC's projections are at their weakest, regarding the future impact of greenhouse gases on global warming. What if life, as a whole, can no longer cope with the changing conditions brought about through global warming? What if life no longer removes carbon dioxide from the atmosphere at the rate that is built into the climate models?

We cannot divorce the study of climate from a study of life and its evolution and, since life is indissolubly part of both its immediate and global environment, we ignore life in the process of climate to our peril.

Perhaps our ancestors did get it right. At least they believed in their effect on the cosmos, hence on the 'order of things'; that it was in their hands to restore equilibrium by returning to some form of natural, accepted behaviour. Are we too late to redress the balance?

Chapter One:
The Nature and History of Climate

The weather, with all its vagaries, is experienced by us every day of our lives. We tend to think of climate as a collection of all these experiences during the course of the year. Yet, as we are aware, different times of the year broadly follow the seasons, which are themselves dictated by the amount of sunlight received during the day. Summer occurs because the Earth faces the Sun for much more of the day than in the winter. Such differences reach their extremes at the poles, which in mid-summer barely see darkness or light in mid-winter. Close to the Equator the length of day remains almost constant. In fact, the Equator receives nearly two and a half times more sunlight during the course of the year than the poles.

Weather shows considerable variation from day to day, let alone from year to year and we need to incorporate such fluctuations into a description of climate. However, if climate is not simply the average of weather, what information do we need to make intelligible pronouncements about its nature? Today's climatologists agree that the minimum period for finding the essence of climate is about 30 years. Consequently, climatologists use a 30-year period to give a reference temperature with which their models must accord, before they can experiment with factors such as increasing greenhouse gas levels. However, even that time span may not be enough to cover the entire range of the natural fluctuations that accompany climate.

At times we can pinpoint the likely cause of short term climatic variation. A volcanic eruption, like that in June 1991 of Mount Pinatubo in the Philippines, can send sun-reflecting particles and aerosols into the upper atmosphere, causing cooler summers for a year or more. Even more dramatic was the eruption in 1815 of Tamboro in the East Indies which blew such a thick veil of volcanic debris into the sky that the Sun barely shone for weeks.

Climate and Weather

If climate reveals regular features that can be depended on from one year to the next, the weather, by contrast, may seem irregular and even chaotic. The temperature shift, within the span of a few hours, may be far greater than the average change in temperature that one would expect from the transition from an ice age to a warm inter-glacial period. In Britain, we describe our climate as temperate, suggesting that we do not experience extremes of climate. Yet, the expressions we use to describe the weather indicate more variability than is suggested in the word temperate. We may experience bitter cold or stifling heat as well as anything in between. In fact, the wide range of fluctuations from day to day, season to season and from one year to the next make it hard to ensure that annual fixtures, such as the Wimbledon tennis tournament, will always take place in good weather. And it is not without reason that farmers of the past attempted to 'make hay while the Sun shone,' since there was always the prospect of continuous rain ruining their crops.

Climate and Land

Mankind has always had an interest in what the weather is doing. For much of the several million years of human existence on the planet, men survived as hunter-gatherers following the movements of animals which chased the seasons. The gathering of plants for food also had a seasonal basis and climate was undoubtedly a major factor in determining what was hunted and gathered, and where. It was probably in pursuit of game that humans first crossed the frozen Bering Straits into the North American continent some tens of thousands of years ago, from where they began the colonization of the Americas.

Evidence suggests that horticulture and the domestication of animals such as the horse, cow, sheep, camel, llama, and the alpaca of South America, took place only after the last ice age had come to an end. Was such an innovation borne of necessity, as more and more bands of hunter-gatherers found themselves pursuing diminishing resources? Or did the end of the ice age herald a milder, more benign climate, allowing hunter-gatherers to stay in one place and begin to experiment with their environment? Farming made long term settlement possible but there was now a double-edge to the weather. On one hand, settlement enabled the storage of surplus produce, so that winters could be faced without having to move away. On the other hand, poor weather during the growing or harvesting season, could lead to disastrous famines and the collapse of civilizations.

Climate and our Past

Before the rise of modern science weather was observed in a descriptive and qualitative fashion. Scribes from ancient Egyptian and Babylonian civilizations recorded harvests, as did medieval monks. The discovery of accounts of almond harvests, as well as wheat and grapes, from ancient cities in the arid Negev Desert, in southern Israel, was actually something of a paradox since such crops require a fair amount of rain. Had the Negev once been wet, and was the desert recent? The area was desert at the time of Abraham some 4000 years ago, and is certainly so today. Recent investigations indicate that more rain fell over the Negev and the surrounding region during the third century BC. Is it a coincidence that cities such as Petra came into being around that time?

It does seem that the local inhabitants, the Nabateans, used their initiative to help both nature and themselves. Some 40 years ago, the Israeli botanist, Michael Evenari, discovered how the Nabateans managed to grow crops in the desert in the centuries before Christ. He found special channels cut into the surrounding hills which converged on specific areas on the plains. He realized that these channels formed a system to collect the run-off from flash floods, and a mere few millimetres of rainfall would provide sufficient water for crops that could sustain surrounding settlements. One such city, Avdat, abandoned for nearly 1500 years, was discovered by the English explorer, T.H. Palmer in the mid-nineteenth century. Using Nabatean techniques, Michael Evenari has transformed parts of the desert into productive horticultural plots that now sustain almonds, wheat, barley and a host of other crops. Not only did he succeed in showing that run-off farming in barren deserts could work, but by making micro run-off catchments he developed the system so that it could be adapted for a flat, featureless landscape.

Prehistoric cave paintings also offer glimpses of what the climate may have been like as long as 40,000 years ago. In his *History of Climate,* H.H. Lamb points out that prehistoric cave paintings in southern France and northern Spain give clues as to the climate of the time, as they depict small bands of hunter-gatherers moving through a tree-less landscape in pursuit of big game, including mammoth, bison, deer, elk, rhinoceros and horses. The ice sheets then covering much of northern Europe were associated with a cold, dry climate that encouraged the growth of savannah and prairies while stunting the growth of woodland and forests. We have evidence from fossilized seeds and vegetation that the climate over much of the planet was much drier than today. Consequently, what is now dense rainforest was then savannah, for example, much of the Amazon Basin. However, a few pockets of land are believed to have kept their humidity and these sources may have provided seeds for the kind of forest we find today. A number of biologists therefore favour the notion that the survival of such *biological refugia* was critical for the conservation of plant and animal species that recolonized the Amazonian rainforest when conditions became more favourable.

The ice, sometimes more than 30 m thick, forced the land mass down: nevertheless, sea level was still about one hundred metres lower than it is today. As a result, humans and animals were able to make their way across corridors between land masses, such as Alaska and Siberia, and Australia and South-East Asia. One area in northern Alaska was probably free of ice on account of it being in the 'rain shadow' of the Rocky mountains, with too little moisture in the air for precipitation. Both animals and man were able to travel south into North America through a corridor that existed between the ice sheets of the Rockies that centred on Hudson Bay.

Better Times

We tend to remember specific years because of unusual weather, like the hot, dry summer of 1976 which was perhaps surpassed by the long dry spell of 1995. Wet summers are also remembered, as are bitterly cold winters. One way we know that climate has shifted, to a warmer or colder state, is by what we can grow. In the early part of the first century BC the Roman commentator, Columella, referred in his work *Concerning Rural Matters (De Re Rustica)* to a report by the 'trustworthy writer, Saserna' that the climate had recently warmed up to such an extent that olives and vines were once again flourishing. The climate also improved in what has been called the Medieval Optimum which lasted from 900 to 1250, and which, according to climatologists was probably, on average, one degree Centigrade warmer than today.

France and England then enjoyed a truly warm spell. Wine pro-

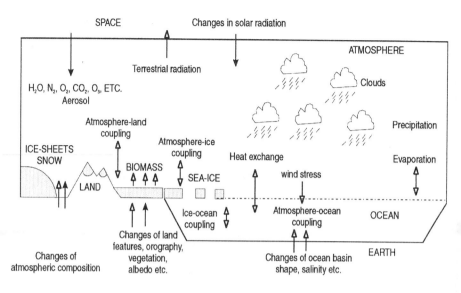

Figure 3. Some of the main components involved in climate. Full arrow heads signify external processes, and open arrows signify internal processes. (Source: Houghton (Ed), The Global Climate, *Cambridge, 1984).*

duction flourished in the south of England and the quality of English medieval wine was such that France tried to have the vineyards shut down, invoking a trade treaty that apparently existed between the two countries. Nor is it coincidental that those times saw the construction of magnificent medieval cathedrals all over Europe; Notre Dame in Paris was begun in 1162 and completed by 1235, while the great cathedral at Chartres was consecrated in 1260 after 66 years of construction. Medieval records indicate that master stonemasons were highly respected and they and their helpers, including the quarrymen, lived well on a diet that included ale as well as considerable quantities of meat. The good harvests of the time and overall prosperity meant that high culture could be afforded; the late twelfth and early thirteenth century was thus a period of epic story-telling, with tales of King Arthur, Tristan and Isolde, Parsifal and in France, the

Roman de la Rose, proving extremely popular. While farmers could grow crops in the uplands of southern England, like Bodmin Moor, where today we can see the outline of their field systems and the relics of medieval houses, now no one would consider arable farming in such a marginal area, which is good for little more than subsidized 'upland' grazing.

Climate and Misery

Climate, therefore, is far from constant; it changes and not only determines what kind of weather we are likely to encounter as the seasons progress, but also whether the planet will warm up or plunge into an ice age. Indeed, the change to a less hospitable climate appears to have come abruptly. The late medieval period was marked by colder weather which was most severe during the seventeenth and eighteenth centuries. The decade 1310–20 was disastrous, especially in Northern Europe. Wet, miserable summers meant that crops failed to ripen and millions of people starved. To add to people's misery the ergot blight, which flourished on unripened, soggy rye, caused St Anthony's Fire; a terrible disease that left its victims suffering from fearful hallucinations and convulsions, to die eventually of gangrene as their limbs rotted and even fell off. Entire communities were wiped out from eating ergot-contaminated rye; some 40,000 dying in Limoges in 943

By the beginning of the fifteenth century, the seas around Greenland had frozen, and the old Viking population died out. Nearly three centuries later the Danish government considered evacuating Iceland because of encroaching sea ice that made access difficult, as well as a massive volcanic eruption that devastated farmland over a wide area. The Thames at London regularly froze over and winters were marked by fairs and people congregating on the ice. Growing food in northern Europe became increasingly difficult and farmers could no longer afford to be taxed on their crops. In Scotland in the 1430s, as in Sweden, successive crop failures led to desperate measures such the baking a type of bread from the bark of trees. Still later, during the seventeenth century, northern Europe was gripped by severe winters, and glaciers again advanced in Iceland, Norway and the Alps, engulfing farmland and forcing entire families off the land. In 1698, Andrew Fletcher of Saltoun in Scotland, spoke of the failed harvests and ensuing famine and death that had struck the upland parishes that year.

Stormy Weather

Storms also became more severe. Sea floods in the thirteenth and fourteenth centuries took a terrible toll, with as many as 330,000 people drowning in one that struck the Dutch and German coasts. In two such floods, in 1240 and 1362, sixty parishes in the province of Schleswig in Denmark were swallowed by the sea, with the loss of half the agricultural land. The island of Heligoland, 50 km in the German Bight, was so eroded by storms that from being 60 km across in 800, it now measures no more than 1.5 km at its widest point. The storms and deaths continued through the seventeenth century and entire communities and towns vanished overnight. On the east coast of Scotland, in Aberdeenshire, all that can be seen of the medieval town of Forvie is a thirty metre high sand dune that covered it during a southerly storm in August 1413.

The changes from one general climatic regime to another, from the Medieval Optimum climate to the Little Ice Age several centuries later, indicate the sensitivity of the system to small changes in the mean surface temperature; changes that globally may amount to little more than one degree centigrade. We can only gain knowledge of the climatic regime through an intensive historical study of past climates combined with collecting data of the prime factors that underlie our contemporary situation. Science, as we know it, emerged from a new way of thinking pioneered by Francis Bacon in England in the first part of the seventeenth century, and by René Descartes in France. Bacon was determined to see nature demystified to make it subservient to man's needs, while Descartes saw nature as a machine that could be unravelled, cog by cog, like an intricate watch. These ideas amongst others inspired the scientists who would reveal the secrets of planetary motions, the nature of the atmosphere, the relationships between living organisms and atmospheric gases, and since the discovery of plate tectonics, the relationship between climate and rocks, continents and volcanoes.

Observing Climate

Astronomy and astrology appear to have been practised by all of the ancient civilizations giving different interpretations to the movement of the planets and stars. In 1543, the Polish astronomer, Nikolaus Copernicus, stated his conviction that the planets moved around the

Sun. Following on from the observations of his teacher, Tycho Brahe, the German astronomer Johannes Kepler developed his three laws of motion and showed that the planetary orbits around the Sun were elliptical rather than circular. Isaac Newton, in the latter part of the seventeenth century, extended the laws of motion to encompass ideas of inertia, whereby bodies in motion would remain in motion and bodies at rest would remain at rest unless a force was applied. He also brought gravity into the equation to describe the planetary motions around the Sun and considerably refined Kepler's work. Newton's description of the forces that hold the planets in their orbits and thus in a long established relationship with the Sun remain applicable today and are used in space research to launch satellites and space craft into desired orbits.

Weather Forecasting

The need for daily, professional weather forecasting followed the sinking of the *Henri IV,* the pride of the French Navy, in a hurricane-force storm on 14 November 1854. On receiving the news, the Emperor Napoleon III, called for weather forecasting services to be established throughout France. A few years later, daily weather forecasts were being published throughout Europe. The United States Weather Bureau was also established at about that time, generating data that would later prove useful in determining any changes in weather patterns during the latter part of the nineteenth century, and the first decades of the twentieth century.

The unpredictability of weather makes weather forecasting a treacherous business, as the weathermen at Britain's Meteorological Office (The Met Office) discovered to their and the country's cost, when, on 15 October 1987, they failed to warn of the approach of one of the most severe wind storms to hit southern England in a century. Weather charts from the previous Sunday indicated the build up of a severe storm in the Atlantic, but later charts used by the Met Office failed to register its strength. Within a few hours of the Met Office's forecast for *unexceptional* weather the next day, storm-force winds, with speeds in excess of 160 kph, caused widespread damage to buildings, and ripped some 15 million trees from the ground. After the event they discovered that the right data had been available but not all the information had been entered into the model in use. A re-run of the data with an improved mathematical model revealed the deep depression of the storm just four hours before it struck London. Prior to the storm, the Met Office had been achieving better forecasts than in previous years and had perhaps become complacent, thinking that the techniques employed were sufficient to anticipate weather conditions for at least twelve hours in advance. History proved otherwise.

Over ten years later the weathermen did get it right and warned of hurricane-force winds that would strike Wales and parts of southwest and southern England on 4 January 1998. Despite prior warning, roofs were blown off buildings, cars were swept away by flooding rivers, trees came down and inevitably some people were killed. That storm followed a succession of storms that continued over several weeks. The appalling weather seemed indicative of some convulsions in climate, possibly attributed to one of the strongest El Niño events of the century and even the past five hundred years.

Putting the study of climate on a scientific footing has been a major undertaking. What has emerged in particular is the extent to, and rapidity with, which climate does change, and how vulnerable living creatures such as ourselves are to such variation. The evidence shows that a moderate climate can alter in a few years to more extreme conditions. Are we responsible for such changes? Or are the forces that bring about such changes more powerful than anything we have done in the past, or are doing now? Two hundred years ago the world population was one sixth of its present size and its overall impact in terms of agricultural and industrial activities was considerably less. During the Medieval Optimum, the population and its impact was smaller again, yet the climate still changed dramatically. We now know far more about external factors that influence climate, including sunspot activity and the periodic variations in the Earth's orbit (see Chapter 2), but we also know the importance of atmospheric gas concentrations on climate, such as carbon dioxide. We now understand that our impact on the concentration of these gases has consequences for climate that may override or augment other relevant external factors.

Desertification

The famous 'Man of the trees,' Richard St Barbe-Baker spent his life dedicated to the restoration of forests. He once told how he could actually watch the Sahara Desert pass over a small object placed on the ground if he was prepared to sit and wait for several hours. He claimed that the desert was expanding by about thirty metres a year on its northern front and a day's observation would show the desert progressing by eight centimetres. St Barbe-Baker firmly believed that the desert could be held back if trees were planted and carefully nurtured during their first years of growth, and he put his ideas into practice in Tunisia.

Further south in the Sahel, those arid lands between the Sahara desert and the lush rainforest of equatorial Africa, rainfall has diminished on average by fifteen per cent since the mid 1960s, from 1200 to just over 800 mm per annum. The decline in rainfall has coincided with years of severe drought, especially 1972–73 as well as 1983–84, in which thousands of livestock and people perished. Robert Mann, who has worked for many years as an agronomist in Gambia as well as other countries of Africa, points out that by 1984 the Sahelian shortfall in rain had lasted 17 years and showed few signs of abating, indicating a substantial change in climate. Mann is convinced that the drying out of West Africa is a consequence of the massive deforestation that has occurred with increasing speed over the past century. By the 1990s more than one million square kilometres of forest and woodland had been cleared along the equatorial zone in West Africa alone. To take one example: in 1939, the volume of wood exported from Ghana was 42,450 m³, by 1987 the amount had risen thirty-fourfold to 1,471,600 m³.

In 1975, the climatologist Jules Charney suggested that deforestation in arid areas of the Tropics would bring subtle changes in the absorption of energy and in wind currents that could lead to a substantial reduction in rainfall. Forests generate humidity through transpiration, whereby water from the soil is drawn up by osmosis through the plant stem and out through the stomata of the leaves. The movement of the moisture and the roughness of the land surface, when trees are present, leads to convection currents in the atmosphere that enhance rainfall. Forests accelerate the movement of water from the soil to the atmosphere and back again, thus keeping the cycle replenished. Experiments with models to mimic such convection processes suggest that rainfall would diminish by as much as one third were forests totally eradicated.

Measurements on the ground also appear to vindicate Charney's hypothesis. Robert Mann cites how, because of the air becoming drier, midday temperatures that used to peak at 35°C are now rising to as much as 65°C. The net result of the increased temperature means even more rapid drying out of soil and of the lower atmosphere, thus instigating desertification. The greater contrast between the temperatures of day and night brings about stronger wind currents and the barren, arid soil is swept upwards into the atmosphere as dust. The quantities of Sahelian dust in the atmosphere have increased significantly over the past few decades. In 1966, meteorologists more than 4,700 km across the Atlantic in Barbados measured six micrograms per cubic metre of African dust: seven years later in 1973, the quantities had increased to 24 micrograms per cubic metre. According to official United Nations figures, desertification is advancing at a rate of 60,000 km² per year out of a total dry land area of 60 million km², of which one third is in Africa.

Shifts in Rainfall

Not all climatologists are convinced that deforestation in Central Africa is responsible for such dramatic changes in rainfall patterns, claiming instead that localized climate changes arise through natural climatic processes, such as changes in solar spot activity and in the movements of the circumpolar vortex. This is a wind system of the northern hemisphere that moves around the Arctic Circle in the depression system generated by the sinking air of the polar cold front meeting the rising air of the warmer mid-latitude circulation system. Shifts in latitude of this jet stream are associated with entire shifts of regional rainfall further to the South, especially over the dry lands of areas such as the Sahel. Climatologists point to even more significant changes in climate that occurred in the past, well before the impact of humans on the environment.

One severely cold period came to an end 14,000 years ago, having lasted approximately one thousand years, and there is evidence of a massive increase in precipitation over the mid-latitude region of North America. Lake Lahontan in Nevada was at least ten times larger then than it is today. To feed such a lake the rainfall must have been greater than the present, and more like the increased rainfall that followed the 1982–83 and 1997–98 El Niños.

History of Climate

Human self-awareness, several millions of years ago, must have undoubtedly brought with it an absorbing interest in climate. Everything, the seasons, sudden storms, the burning Sun, would have needed an explanation. But, given our long multimillion year history, it is surprising how recently we first began to apply a scientific rather than a pantheistic model of climate. Although, before Plato, the Greeks believed in a heliocentric universe, that view changed with Plato and in the second century ad the Egyptian astronomer, Claudius Ptolemaeus, had the Sun moving around the Earth. Ptolemy's mistakes later served the Christian church well and theologians, at least until the Renaissance, were delighted to have proof of the Earth's primacy at the centre of the universe. In 1543, disrupting fanatically-held religious beliefs, the Polish astronomer, Nikolaus Copernicus, published his *De revolutionibus orbium coelestium* demonstrating that the orbit of the planets, and most significantly that of the Earth, could be explained better mathematically if they all revolved around the Sun. Three-quarters of a century later, in 1616, the Inquisition reacted by condemning Copernicus' work and all who believed in it. No one was exempt, not even a scientist of Galileo's calibre, and after he denounced the Ptolemaic system of astronomy in defence of Copernicus in his 1632 book, *Dialogue on Two World Systems,* he found himself arraigned in front of the Inquisition and forced to retract on pain of death. As it was, he was confined to his house in Florence until his death ten years later. It would be wrong to conclude by that that Galileo did not believe in an all-powerful creator; he did, as did Newton, born in 1642, the year Galileo died. But the times were certainly revolutionary, and René Descartes (1596–1650), was perhaps the first in history to propose that the Earth, the Sun and all the planets arose from a single process. This *nebula hypothesis* is still in vogue today, with hard science to back it up.

Galileo, Torricelli and Climatic Instruments

Science often proceeds with the invention of an instrument which then has applications over a wide range of investigations. We take it for granted now that to have any idea of climate we need to measure air temperature, humidity and pressure, as well as the direction in which the wind is blowing. Early in the seventeenth century Evangelista Torricelli invented the barometer. The first barometer consisted of a tube of water closed at one end which was inverted with the open end in a bowl of water. Changes in the pressure of the atmosphere were registered by changes in the height of the water column. The actual atmospheric pressure raised a column of water nearly one metre. Torricelli soon realized that a heavier liquid was easier to manage and replaced the water with mercury which, with a specific gravity 13 times heavier than water, needed a much shorter glass tube. With its silvery colour, it could also be seen more easily.

Galileo was a contemporary of Torricelli, and is renowned for his proposition that objects of different masses, such as feathers and lead weights, would fall to Earth equally fast were the planet airless and offered no resistance to a falling object. In 1609, Galileo developed his own telescope and discovered four satellites of Jupiter as well as the reason why the Moon shone. He was also known for his invention of the thermometer, mercury again providing the best liquid since it had a higher relative expansion on heating compared with water.

Wind, perhaps, was the most tangible of phenomena since it could be observed making trees sway and agitating the sea and not surprisingly the weathervane was invented centuries before the other instruments. Accurate ways of measuring wind with revolving cups were to come only in the twentieth century, yet even these have their limitations, working best for wind speeds that exceed 5m/s. For lower wind speeds, meteorologists use thermal anemometers which measure the degree to which flowing air cools tiny, red-hot wires.

Physics of the Atmosphere

The physics of the atmosphere followed the invention of the thermometer and barometer. Robert Boyle, a seventeenth century physicist, showed that the volume of a gas at a constant temperature is inversely proportional to its pressure. The Frenchman, J.A.C. Charles, later showed a similar inverse relationship between temperature and volume, when pressure is constant. Boyle's Law and Charles' Law have since been put together in a composite law which implies that the three qualities of volume, pressure and temperature when relating to a gas are dependent upon each other. This composite law is essential in understanding what happens to masses of air in the atmosphere. As a result of air's compressibility, the lower layers of the atmosphere are much denser than those above. Even though the atmosphere extends upwards beyond 80 km, fifty per cent of the total mass of air is found below 5 km, at which altitude (16,000 feet) its average density has dropped by nearly one half compared to its density at sea level.

Historical Records

The scientists of the seventeenth century gave us the basis for the study of climate. They could make reasonably precise predictions regarding the movement of the Earth around the Sun and could measure, albeit crudely, some of the essential attributes of climate and weather. Various efforts have been made to unravel the climatic history of the past few centuries based on early scientific records. After painstaking work to compare different records from overlapping sites in central England, the meteorologist Gordon Manley obtained a series that went back as far as 1659. The UK Meteorological Office, at the Hadley Centre, has updated these records which provide some of the best historical information we have. Others have carried out equivalent studies in countries such as Norway, Holland and the eastern United States. These historians are in general agreement that temperature data up to 1720 is accurate to within 1°C. With the development of better apparatus and a more careful approach to data collection, the information then improves and is accurate, it would seem, to 0.2°C. Climatologists have had greater difficulty getting trustworthy data on rainfall in past centuries owing to serious doubts about the representativeness of the sites. Samuel Pepys was one seventeenth century source of information, and Thomas Barker of central England made daily weather observations that ran from 1733–98, noting in particular any change in the weather, especially increased rainfall, that began in the 1740s.

The Reverend William Merle was one of the first Europeans to record the weather, between 1337 and 1344. 250 years later the Danish astronomer, Tycho Brahe, also carried out weather observations, which like Merle's, proved reasonably accurate and a good reference for those studying the history of climate. Modern historians can find clues about past climates by sifting through records of harvests and assessing not just the size of the harvest, but precisely when in the year it was completed.

Times have changed and, instead of a few amateurs recording rainfall in a handful of sites, today reliable measurements are taken on a daily basis at thousands of representative sites. Mike Hulme of the University of East Anglia has reviewed data collected from some 8,300 different sites across the globe and has found evidence in recent years of significant increases of rainfall in northern latitudes above 50° with corresponding drops over the Tropics. The 1980s proved to be both the warmest and the wettest in high northern latitudes, at least from records going back over the past century.

Climate Models

The elaboration of General Circulation Models (GCMs) for the study of climate dates back to the 1920s and the work of the Englishman, Lewis Fry Richardson. After several months of calculation using a battery of slide rules with which he hoped to account for the movement of masses of air, involving temperatures, pressures and humidities, Richardson attempted to predict what the weather would be like 24 hours after his starting time. As it happened, his results proved to be wrong. Today's computers, combined with a vast range of constantly updated information, gleaned from satellites as well as ground and ocean-based monitors, are likely to be reasonably accurate in the short to medium term. Currently, the most advanced of such computer systems is at the European Centre for Meteorology at Reading, which every day can *number crunch* a million weather observations at the rate of 250 billion calculations a second. The centre claims it can forecast the world's weather up to ten days ahead with eighty per cent accuracy for five days.

Chapter Two: Sun, Earth and Moon

Driving Forces

The Sun is the driving force behind the Earth's climate, bathing our planet in highly energetic electromagnetic radiation, including visible light, ultraviolet light and heat. The Sun has a mean temperature of around 6000°C which manifests itself as a continuous stream of photons pouring into space. In terms of energy these photons amount to 73.5 million watts per square metre (Wm^2) , of which at any one time the Earth intercepts just five thousandths of a per cent, equivalent to each square metre of the Earth's outer atmosphere being heated by a small 340 W heater. 45 per cent of this radiant energy is in the visible part of the energy spectrum and is therefore revealed as light. A similar proportion appears as heat in the infra-red range, and just nine per cent is ultraviolet.

Terrestrial life is fortunate that ozone, during its generation from oxygen, reduces the effects of the most energetic and damaging ultraviolet rays, known as ultraviolet C (UV–C). Once created, ozone interacts with ultraviolet B (UV–B), those rays with intermediate energy between the powerful UV–C and the feeblest ultraviolet A (UV–A). Virtually all the UV–C passing through the stratosphere is transformed to less energetic electromagnetic radiation because of the interaction with oxygen as does most of the UV–B, but this time with ozone. The remaining UV–A tends to stream, unimpeded, through to the surface of our planet.

The Tilt of the Earth

The Earth was once spinning much faster, but has now settled into a motion that takes it close to 24 hours to make one complete turn. Several billion years ago the Earth may have spun five times faster, which would have had profound effects on the movement of air in the atmosphere; in particular it would have inhibited air from the equatorial, tropical zone from reaching the poles. As a result the

Tropics would have been warmer and the poles cooler as well as drier; life at these times would have been limited to the seas and shoreline. By the middle Devonian times, 390 million years ago, the spin had noticeably slowed and, by studying the growth rhythms in fossilized corals, the biologist John Wells came to the conclusion that by then a year was down to 400 days.

Fortunately for us the Earth is presently tilted from the perpendicular by 23.44, resulting in our seasons as first one hemisphere leans towards the Sun and then the other six months later. Without the seasons — if the Earth were spinning upright — the polar regions would never receive extra summer Sun to keep the ice at bay, and we would probably be locked into a perpetual ice age. Not that the tilt is fixed; it fluctuates over a regular period or around 40,000 years between an angle of 21.8 and 24.4 and is currently decreasing by about one ten-thousandth of a degree every year. The more the Earth is tilted, the more marked the difference between the seasons. And, it is because of the tilt that the equinoxes occur, when the Tropics of Cancer and Capricorn respectively face the Sun.

However, that is not all: added to the asymmetric tilt is the actual course the Earth takes on its annual journey around the Sun. Over a period of 90,000–100,000 years, owing to the alignments of our neighbouring planets and variations in their combined gravitational pull, the Earth shifts its course from one that is circular to one that is more elliptical. When the path is nearly circular the amount of light and energy reaching the Earth over the year is more or less constant (apart from phenomena such as solar flares and changes in sunspot activity). When the path is truly elliptical, with the Earth first relatively close to the Sun and then further away, the solar energy received by the Earth may vary by as much as thirty per cent over the course of the year. Such variations have a strong impact on climate, with each hemisphere the mirror image of the other. Thus, when one hemisphere is closest to the Sun during its summer, the other will be having its winter and vice versa. The Earth is now experiencing a slightly elliptical orbit; as a result, those of us in the northern hemisphere receive 3.5 per cent more energy in the winter compared to what we would receive if the orbit were circular, and about 3.5 per cent less in the summer.

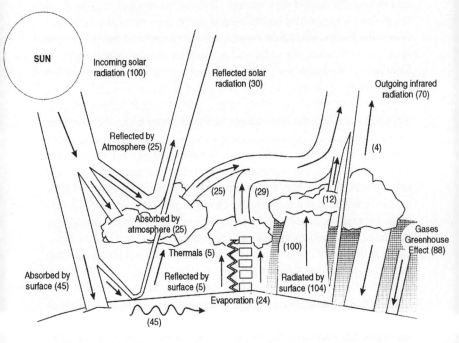

Figure 4. Earth's radiation energy balance. The incoming radiation of 340 Wm² is taken as 100 per cent. 88 per cent of the absorbed energy is re-directed to the Earth's surface as a consequence of the greenhouse gases. (Source: Schneider, Scientific American, *1987, 256:5, 72–80).*

Axial Precession

To confuse the situation further, which hemisphere is closest to the Sun during a particular season also varies according to a cycle of 25,000 years. 12,000 years ago, when we were coming out of the last ice age and going through a period of comparative warmth, the northern hemisphere enjoyed its summer when it was closest to the Sun and its winter when furthest away — the antithesis of today. At that time the northern hemisphere received 5–10 per cent more sunlight during the summer than it does today, and a similar proportion less during the winter. This process of *axial precession* results from the North Pole, actually wobbling around the axis of a line drawn between the North and South Poles. The wobble rotates the poles, and thus the Earth itself, in a clockwise direction, which is the opposite direction to the Earth's spin. If we were alive 13,000 years hence we would not be looking to Polaris as the North Star, but to alpha-Lyrae on the opposite side of the sky compared to where we look today. This has considerable consequences for climate, since the northern hemisphere has far more continents and a greater land mass compared to the southern hemisphere. Land masses tend to respond more quickly to heat input than the ocean; the land absorbs and sheds heat from one day to the next, whereas the oceans retain their heat for longer, thus balancing out changes in solar input. In past history, ice ages seem to be associated with circular rather than elliptical orbits and with less tilt in the Earth's axis.

Milankovich's Wobble

In the 1920s Milutin Milankovich, a Serbian meteorologist, further refined the idea that tilt and orbital changes during the course of millennia were sufficient to affect climate. But confirmation that ice ages over the past few million years are caused by the *Milankovich Wobble* has been hard to come by, despite the conviction of some climatologists that they can see correlations between the 100,000 year shift in the orbit around the Sun and the coming and going of ice ages. Perhaps a combination of factors is involved, with Milankovich's Wobble giving the impetus that switches from one climatic regime to another.

The Moon

The Moon is another crucial factor affecting climate and the ancients were perhaps wise to pay as much respect to the silvery planet as to the Sun. Without the Moon, the Earth would have no tides, but far worse, its rotation would be unstable and a Moon-less Earth would probably tumble chaotically around the Sun, leaving little possibility for seasons and no semblance of a stable climate. As the French mathematician Jacques Laskar demonstrated in 1993, without the Moon's calming influence the Earth's tilt could vary between 0° and 80°. The amount and intensity of sunlight that strikes the Earth's surface is also affected by the movement of the Moon around the Earth. The Moon pulls the Earth ever so slightly closer and then further away from the Sun, therefore marginally but significantly changing the amount of solar radiation reaching the Earth. In all probability, an Earth without the Moon to stabilize it, so that we have distinct seasons, would not have been hospitable to life.

Reflection

On average about one quarter of the Earth's outer surface is exposed to the Sun at any one moment, yet not all the 342 Wm^2 of energy actually gets right down to sea level. Clouds reflect roughly thirty per cent of incoming radiation back into space: the remaining seventy per cent, 240 Wm^2 heats the Earth's surface and atmosphere. The shininess of the Earth's surface and atmosphere makes a fundamental difference too, since the whiter, or more glassy the surface the more sunlight is reflected back into space. We know that light reflects off snow and ice — that is why we get sunburnt skiing and have to watch out for snow-blindness — and even light-coloured deserts tend to reflect more light back into space than they absorb. In contrast dark-coloured surfaces usually absorb light and heat up, which is why we use black plastic to propagate seedlings, and use a dark or black background in flat-bed solar water heaters to maximize the amount of heat we can trap. If clouds tend to cool the Earth, dark oceans under a clear sky absorb light and get warm.

The Earth's climate therefore results from a combination of the Earth's motion around the Sun and features of the Earth's atmosphere and surface which either reflect or absorb light and heat. Since the Sun shines down on a spinning, off-centre sphere — our planet — its rays are perpendicular to the surface only in one spot during one moment in time. That constant disparity helps generate climate by generating gigantic convection cells. Cooler air from one region is drawn to a warmer region, while warmer air compensates by moving to a cooler region. Consequently, energy is transferred from one part of the planet to the other. The French physicist, Gustave de Coriolis, early in the nineteenth century, first realized that the Earth's rotation could pull the air masses away from the central plane and especially near the poles, where the distortion of the air currents would be at their maximum. It is this force which gives us the south-westerlies and the Gulf Stream in the northern hemisphere, and the corresponding north-easterlies and ocean currents in the southern hemisphere. But, for that to happen, the rate of spin has to be more or less as it is today.

Ice Ages

In 1837, the palaeontologist and geologist, Louis Agassiz, after climbing mountains in his beloved Switzerland, came to the conclusion that the scourings of the rock face and the movement, even uphill, of massive boulders in the Jura Mountains must be evidence that massive glaciers had once covered much of northern Europe and transformed the landscape into its peaks and valleys. Once geologists knew what to look for a spate of findings elsewhere, unravelled similar features across the North American continent and the rest of Europe, including Britain. Glaciers, it seemed, had once stretched for at least 2,300 kilometres between the 50th and 72th parallel. The discovery that the planet had undergone at least one ice age was something of a shock, and reasons had to be found.

Carbon Dating

A French scientist, Joseph Alphonse Adhémar, was one of the first to speculate whether variations in the Earth's orbit around the Sun might have something to do with a changing climate. 20 years later, in 1864, James Croll, a Scottish geologist, put some numbers to this idea and was convinced that even small shifts in the Earth's orbit would be sufficient to bring about climate change, including swings between ice ages and warmer times.

Although Croll made some intelligent guesses as to when the last ice age occurred, accurate dating had to wait until the American chemist Willard Frank Willy tumbled on the idea in the 1950s of dating fossilized remains, including vegetation and animals, using radioactive substances which decay at set rates. Since carbon is the basis of organic chemistry and is drawn out of the atmosphere as carbon dioxide into living organisms, Willy suggested that radioactive carbon–14, a natural component of carbon dioxide, could offer a relatively precise method of dating back to the time when the organism was last living, by noting how much of the radioactive carbon had decayed. A particular mass of carbon–14 loses one half of carbon–14 in 5,730 years. The measurement of radioactive decay is in half-lives: a given mass of carbon–14 decays to half its mass over one half-life; to one quarter during two half-lives; and to one eighth during three half lives.

After 100,000 years very little of the original carbon–14 is left, thus giving a natural limit to the use of carbon–14 for telling us how old a fossil is.

Carbon-14, Solar Flares, Sunspots and the Aurora Borealis

As for carbon–14, given that it is always undergoing radioactive decay, why hasn't it all disappeared? The reason is that it is created from atmospheric nitrogen, and over time, the amount of carbon–14 being generated in the upper atmosphere and the amount vanishing through radioactive decay reaches some kind of equilibrium. But it is not as simple as all that. The amount of carbon–14 created does differ slightly from year to year, mainly because of variations in the streaming of high energy particles from the Sun. This *solar wind* actually interacts with the cosmic rays; the stronger the solar wind the less interaction of the cosmic rays with nitrogen. Consequently during a time of solar flares, when the solar wind is stronger, the amount of carbon–14 tends to decline.

This discovery gave Minze Stuiver of the University of Washington in Seattle a way of determining times of solar flares over the past thousand years and of estimating the proportion of atmospheric carbon–14 from year to year. By analysing the carbon–14 content in annual tree rings of species such as the bristlecone pine, Stuiver and his colleague, P.D. Quay, were able to pinpoint which years were abnormal in the sense of having higher or lower concentrations of carbon–14 compared to a control. For instance, in 1000 the amount of carbon–14 was about minus five per thousand compared to 1890. Five hundred years later the relative amount of carbon–14 was up eight per thousand compared to 1890. The quantity of carbon–14 increases in the bristlecone pine when solar activity is relatively low, and decreases during solar flares.

When the Earth's surface temperature is matched against such ups and downs an inverse relationship emerges. The relatively low levels of carbon–14 in tree rings 900 years ago, during the so-called Medieval Optimum, correlate well with the warmer climate of that time, when average temperatures may have been as much as 1°C higher than today. And what about the periods of cold weather that lasted for several hundred years, until the early nineteenth century? Again the link between high relative levels of carbon–14 and a colder climate seems to be borne out.

Solar flares accompany *sunspots* and some climatologists maintain that there is a connection between sunspot activity and climate. John A. Eddy, of the High Altitude Observatory in Boulder, Colorado, has spent more than twenty years investigating the association between sunspots and global temperatures. As he points out, people have been looking at sunspots for millennia; the Chinese were doing so as far back as 26 BC, and between then and the sixteenth century they recorded 150 sunspots, about one every decade, and all before the invention of the telescope. The ones they would have seen were large since we now have counts of 180 and more during times of enhanced activity. Enhanced solar activity also causes spectacular displays over the polar skies, brought about by interactions of high energy particles and atoms, such as the *aurora borealis*, which people have observed for centuries, if not millennia. When we compare the results of radiocarbon in tree rings with observations of the aurora we find a close correlation; the aurora highlighting the skies when sunspot activity is increased, and vanishing during the quiet periods of the Sun's history. Nevertheless, it was the telescope which transformed our ability to observe the Sun, so that when we combine data on sunspots with changes in carbon–14 in the annual rings of trees, we begin to see close correlations and corroborative evidence for climate change in the past.

Observations of sunspot activity over the past 120 years show that the numbers wax and wane over an eleven-year cycle. For instance, peaks in 1969 and 1979 were preceded by troughs in 1964 and 1976. But, the peaks are not uniform and an uneven upward trend between 1870 and 1960 is discernible, with a substantial threefold increase in the number of sunspots becoming apparent in the latter year. Trends since have been down again. Meanwhile, through the use of satellites we can now measure solar radiative flux; the Sun's energy at the top of the Earth's atmosphere. Here again, correspondence has been found with sunspot activity and solar flares, as well as with a colourful display of the aurora borealis. Most of the Sun's energy comes to us in the visible spectrum, but some comes as short wave ultraviolet rays and x-rays. It is in this short wave region in particular where we detect most of the increases associated with solar flares and sunspot activity. In fact, the ultraviolet reaching the Earth nearly tripled in 1960 compared with the end of the last century.

From 1100–1300, we know that Europe experienced an uncommonly warm climate with incredible consequences for culture. Indeed, historians of the Middle Ages called the thirteenth century 'the greatest of all centuries,' when magnificent cathedrals and works of art flourished. It was also a time of successful agriculture, which combined with the newly-invented collar and harness for horses, led to the opening up of new lands, and the doubling of the European population. We have evidence that solar activity, including solar flares and sunspots, may have been double that of today, and the obvious question is whether enhanced activity accompanies a warmer climate?

However, if the Middle Ages were warm, the next couple of centuries, with some respite in the late fourteenth century, were not, and Europe suffered bitter cold and miserable weather that destroyed harvests, particularly in the North. Millions died of starvation, if not of the plague. Again we find a correlation between the chill period of the Little Ice Age, from about 1650 until 1720, and what appears to be a minimum of sunspots. T. Wigley and M. Kelly, of the Climatic Studies Unit of the University of East Anglia, looked at variations in solar energy for that period and compared them with radiocarbon data. From a simple climate model they then concluded that the Little Ice Age could have resulted from a quarter to one half per cent decline in solar energy.

Despite such correlations and climate models, climatologists are divided in their opinions as to whether sunspot activity is associated with global warming or cooling. Some believe that variations in solar activity are a nearly complete explanation for changes in climate; certainly more important than changes in greenhouse gas concentrations. Others believe that the emphasis for climate change has to be put on greenhouse gas emissions, especially since physics shows these gases to be extraordinarily potent in raising temperatures, even at relatively low concentrations.

Sunspots and Cooling

The Sun's output varies slightly from day to day, with fluctuations that for the most part flicker around a few tenths of one per cent of what meteorologists recognize as the *solar constant*. Variations in the Sun's output are now associated with sunspot activity.

Sunspots indicate cool regions on the Sun's surface but, paradoxically, the more there are the greater the solar output. For instance, sunspot activity appears to have nearly died away from 1645–1715, a period which was marked by prevalent low temperatures across the planet and freezing winter weather. However, astronomers have now discovered that sunspot activity is associated with an increase in bright areas, known as *faculae*, around the sunspots, which compensate for the diminished electromagnetic radiation from the sunspots' interior. Sunspot activity has a somewhat variable cycle of just over eleven years in which the numbers go up and down. Shorter cycles of about nine years are often associated with greater sunspot activity and these periods coincide with warmer surface temperatures, such as late Roman times and the Middle Ages. By the same token, periods of cool surface temperatures, such as 1400–1510, a period known as the Spörer Minimum, and the Maunder Minimum of the seventeenth century, coincided with low sunspot activity. During the Maunder Minimum, the sunspot cycle lengthened to 23 years and the Sun's brightness faded by at least 0.4 per cent.

Is it possible that past climates were influenced by sunspot activity? Historical records indicate that, during the Medieval Optimum of the twelfth and early thirteenth centuries, solar activity may have been double today's. Furthermore, increased solar activity is associated with more UV–A and UV–B reaching the Earth's surface, particularly in the Tropics where the ozone layer is thinnest. A review of the age-adjusted incidence of malignant melanoma in Connecticut between 1935–75 shows an upward trend that probably reflects the increased time that modern humans spend outdoors, exposed to the Sun. However, superimposed on the upward trend are peaks of varying size that closely follow the upside of the sunspot cycle.

Some climatologists even wonder whether the shift in the solar constant brought on by sunspot activity may even have led to the demise of civilizations. John Eddy, of the International Earth Science Information Network, questions whether the sudden abandonment and collapse of one cultural centre after another in Mesoamerica, at the time when Europeans were flourishing under the warm Sun of the early Middle Ages, may have been the result of intolerable exposure to UV–A in the Tropics.

The argument regarding the impact of solar output on the ups-and-downs of surface temperature has been brought up to date by the work of Eigel Friis-Christensen and Knud Lassen of the Danish Meteorological Institute. They looked at surface temperatures across the planet over the past 250 years and compared them with the length of the solar cycle, the inference being that the shorter the cycle the higher the output from the Sun. The correlations have held fast for this century and the sunspot cycle is now months shorter than it was one century ago. However, more important is the number of sunspots in evidence at any one time. Here, we have a discrepancy — the number of sunspots has gone down since 1960 — an indication that the Earth should be getting cooler, at least on the surface. That does not fit with the rapid warming experienced over the past couple of decades, all of which indicates that some other factor must be operating.

Friis-Christensen and Lassen have returned to the fray with ideas that cosmic rays penetrating the Earth's atmosphere cause clouds to form that tend to cool the planet. Since the increase in the solar wind that accompanies increased solar activity interacts with cosmic rays and deflects them, less clouds should form and leading to more heating. Yet again, the evidence for such a correlation is suspect, especially since other factors such as volcanic eruptions and El Niño have a more obvious and powerful effect on tropical cloud formation. If, as shown above, sunspot activity has been in decline since the 1960s, that again would be evidence that the changes we are experiencing are primarily a result of the rise in greenhouse gases.

In conclusion, before the industrial revolution any flux in the solar constant, especially in the proportion of energy in the ultraviolet part of the spectrum, would be bound to have some impact on global temperatures and could be responsible for such phenomena as the Medieval Optimum and by contrast the cold that swept in during the Maunder Minimum. However, after World War II, when industrial development took off with a vengeance, any changes in solar constant would have been outweighed by changes to the atmosphere in terms of greenhouse gases and cloud formation. On the basis of greenhouse gas emissions alone, climatologists' general circulation models (GCMs) indicate that we should expect average temperature rises of 4°C or more over the next century. That would mean an average 0.4°C rise every decade at the very least; an enormous increase unprecedented in human history.

Chapter Three: Ice Ages and Greenhouse Gases

Unquestionably, climate has undergone dramatic changes during the history of the Earth. One hundred million years ago, when the dinosaurs were still in existence, the planet appears to have had no permanent ice; all the evidence points to a warmer planet than now, with a tropical climate stretching from the Equator to the poles. Seas were not only one hundred metres higher than today, but at high latitudes some 15°C hotter. Then, four million years ago when humans were evolving in East Africa, the climate began to change dramatically and the Earth not only began to have permanent ice at the poles, but underwent periodic cycles between one glacial period and the next. During ice ages ice stretched south from the Arctic Circle. In interglacials, most of the ice sheets melted. The cycles have changed in length as time has passed. Several million years ago they lasted 40,000 years; more recent cycles appear to have lasted 100,000 years, with the glacial part of the period taking up nine tenths of the whole. Interglacials have therefore been lasting little more than ten thousand years. Since we are now 10,000 years on from the last ice age, we are definitely due for another, but perhaps we have put paid to that with global warming? At the peak of the last ice age 20,000 years ago, giant ice sheets, heaped into domes more than a kilometre thick, stretched across North America from New York State to the Rockies, and across equivalent latitudes in northern Europe.

Albedo and Positive Feedback

The abruptness of their melting is a phenomenon that is not properly understood, but it clearly entails a process which, once begun, proceeds at an accelerating pace until most of the ice has gone. As an ice sheet shrinks it exposes more and more ground to sunlight. In contrast to ice and snow, both of which reflect sunlight away, exposed rocks and soils absorb heat. The degree to which a surface re-

flects light from the Earth is known as *albedo*; a highly reflecting
surface, such as fresh snow, has an albedo of around ninety per cent,
and the oceans, which absorb heat and light, have an albedo of
roughly ten per cent. As ice and snow melt, exposing ground, so
more heat is absorbed, which in turn leads to more melting and the
exposure of more ground. This snowball effect is categorized as *positive feedback*. The same process can, of course, go in exactly the opposite direction, when enhanced cooling leads to the spread of ice
and snow and consequently to an increasing proportion of sunlight
being reflected away.

Thus, the higher the overall albedo of the Earth's surface the
greater the chance of the planet entering an ice age, and the lower the
albedo the greater the chance of an interglacial. Other forces have
therefore to come into play to move the planet out of a glacial or interglacial period. The mean temperature difference between the two
states is considerable and the evidence demonstrates that the Earth's
surface temperature is, on average, now 5°C warmer than it was
20,000 years ago. The previous interglacial, known as the Eemian,
which ended 114,000 years ago, was even warmer than the current
interglacial; hippopotamuses lived among the Thames marshes, and
elephants and lions roamed in Cornwall and the south-west of England. Meanwhile, most of the West Antarctic ice sheet had melted,
sea ice was only found in the Arctic Ocean, and sea levels were 6–
8 metres higher than at present. Nevertheless, any impression of a
steady, warm comfortable climate would be mistaken. Ice core samples from Greenland indicate that temperatures fluctuated between
warmer and cooler spells during the early part of the Eemian, with
sudden drops of as much as 10°C putting the planet back in the grip
of bitter cold.

Even though the ice during the glacial periods was limited to the
upper latitudes of the northern hemisphere, the consequences
around the globe were considerable, with the level of the sea more
than one hundred metres lower and a much drier climate in the
lower latitudes. Savannah rather than rainforest covered much of
the Amazon Basin, and the Sahara Desert stretched several hundreds of miles further south than it does today, as evidenced by fossil sand dunes in the region known as the Sahel. By 6000 BC, owing
to a warmer, wetter climate, Sahelian vegetation had advanced 5°
to 21° North, only to retreat once again as the climate dried out.
The assumption is that such shifts from dry to wet and back were
due to natural causes, including fundamental changes in prevailing

wind patterns, and could result from shifts in the amount of energy reaching the Earth's surface from the Sun, as well as from orbital changes, such as Milankovitch's Wobble.

The periodic switch between ice ages and interglacials, over the past two million years, has led some climatologists to believe that the causes of such dramatic climate shifts would completely obscure the climatic consequences of human activities. If such sceptics of human-induced climate change were correct, then the political consequences would be profound, since industries and other major emitters of greenhouse gases, such as carbon dioxide, would demand to be let off the hook. However, the evidence now points to human activities, resulting in the atmospheric accumulation of greenhouse gases, as being largely responsible for climate change and for more extreme climatic events, such as strong and persistent El Niños.

Greenhouse Effect and Clouds

How much sunlight gets through to the Earth's surface and how much is reflected away is critical to the Earth's energy budget. However, there is another factor, of crucial importance, which has to do with the retention of heat through the action of greenhouse gases in the atmosphere. Although rarely mentioned in the same breath as carbon dioxide, the most significant greenhouse gas is water vapour, which accounts for up to seventy per cent of greenhouse gas warming. Without the initial background warmth provided by carbon dioxide and methane, as well as other greenhouse gases, water would barely vaporize in the first place. This is a classic self-generating feedback in which the more warmth in the atmosphere the more water evaporates from the surface, and the more water vapour in the atmosphere the more the atmosphere warms. However, this situation cannot go on forever and a point is finally reached when the dynamics of water evaporation, followed by precipitation, cancel each other out. On the other hand, should global warming occur then the amount of water vapour held in the atmosphere also increases, thus accentuating warming. Climatologists believe that, if a doubling of carbon dioxide from pre-industrial levels were to raise surface temperatures by 1.2°C, the water vapour then drawn into the atmosphere would amplify the temperature increase to 1.9°C.

If clouds are brought into the equation a complicated picture unfolds. Clouds have a twofold effect: one is to reflect light away and so help cool the surface which appears to be the dominant effect; the other is to hold back heat radiating up from the Earth's surface. Climatologists estimate that on average clouds reflect away 44 Wm² of sunlight and their whole greenhouse effect is equivalent to 31 Wm²; the net balance is therefore a cooling of approximately 13 Wm².

But what happens following global warming? Earlier models indicated that cloud cover would decline and since the light-reflecting nature of clouds dominated over heat retention, the assumption was that the change in clouds could bring about additional warming. More recent modelling indicates that a warmer climate might lead to more clouds which would therefore offset some of the warming through light reflection. On the other hand, warmer clouds tend to rise higher and a higher cloud is a colder cloud; hence, its ability to emit heat out to space is reduced. One result indicates that an increase in greenhouse gases could reduce the overall cooling effect of clouds by as much as one sixth — from 13 watts per square metre to 11 watts. Just that decline in cloud-cooling would increase by half the global warming effect resulting from a doubling in carbon dioxide concentration.

Greenhouse Gases and Global Warming

Jean-Baptiste Fourier was the first to suggest in 1827 that the Earth's atmosphere behaved like glass in a greenhouse which let the Sun's energy in, as light, but held back heat and therefore caused the atmosphere to heat up. In 1860, John Tyndall measured the extent to which carbon dioxide and water vapour absorbed long wave radiation. Astonished at the apparent power of the greenhouse gases to heat up the atmosphere, Tyndall argued that fluctuations in atmospheric concentrations of carbon dioxide might be responsible for recently discovered changes in climate between ice ages and interglacials. In 1896, Svante Arrhenius concluded that doubling the atmospheric carbon dioxide concentration, as it was then, would lead to a global warming of as much as 6°C. Fifty years later, in 1940, the British physicist G.S. Callendar corroborated Arrhenius' calculations by estimating global warming brought about through carbon dioxide emissions from burning fossil fuel.

Typically, carbon dioxide (CO_2), water vapour (H_2O), methane (CH_4) and nitrous oxide (N_2O) have the potential to resonate with infra-red photons, and since each molecular species has its own resonant frequency, they cover a considerable range of the infra-red spectrum. Satellites orbiting above the atmosphere indicate the extent to which the different gases absorb thermal radiation and at what frequencies. Certain frequencies are unaffected by the presence of greenhouse gases, and they pass through the atmosphere without hindrance. Altogether, the different greenhouse gases complement each other, covering a good part of the full infra-red spectrum. The extent to which the atmosphere warms up depends on their concentrations at least up to the point of saturation, beyond which the infra-red of a particular wavelength passes out to space relatively unhampered.

The year 1957 was a moment of profound significance for global climate studies. It was International Geophysical Year and the United States offered to establish an atmospheric laboratory on Mauna Loa, an extinct volcano in Hawaii. It was a perfect site, high up on the mountain, in the middle of the northern Pacific Ocean and therefore away from the big continents and far enough away from local industrial emissions. Measurements of atmospheric gases and aerosols were therefore representative of global concentrations of greenhouse gases in the atmosphere, as mixing takes place within a few months on each side of the Equator. Charles Keeling, the laboratory's director, had already made a name for himself with his studies of the carbon cycle. His pioneering work meant that the origin of recent atmospheric carbon emissions could be identified, from the burning of fossil fuels and wood, and from other non-fossilized organic material.

Carbon Dating

Carbon exists in a variety of isotopes of which just three are important for studying the natural world: carbon–12, the most common; carbon–13, which like carbon–12 is stable; and finally carbon–14, which is radioactive. The half-life of carbon–14 gives the means to find out the age of dead and fossilized organic matter. Since fossil fuel has long since lost carbon–14 through radioactive decay, it explains the two per cent fall in the concentration of carbon–14 in the atmosphere that occurred between 1800–1950 as measured through the uptake of carbon–14 into trees during that period. This tallies with carbon dioxide emissions since the industrial revolution began two and half centuries ago.

In fact, between 1850 and 1986, the burning of fossil fuels released nearly 200 gigatonnes of carbon into the atmosphere and deforestation to convert land for agriculture added another 120 gigatonnes. The current concentrations of carbon dioxide indicate

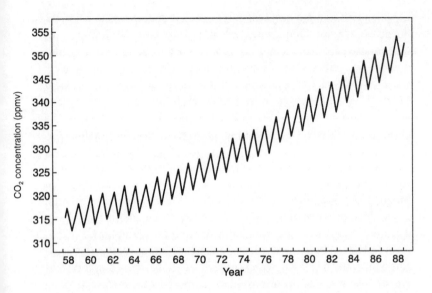

Figure 5. Monthly average carbon dioxide concentrations in parts per million of dry air, observed continuosly at Mauna Loa, Hawaii. The seasonal variations are due primarily to the withdrawal and production of carbon dioxide by the terrestrial biota.

that almost half of these emissions are still in the atmosphere. For the past decade, we have been emitting 7.5 gigatonnes of carbon in the form of carbon dioxide each year, an amount which is close to one percent of total atmospheric carbon dioxide. We must conclude from historical and scientific evidence that the net accumulation of carbon dioxide in the atmosphere since the industrial era began is a direct consequence of fossil fuel burning and deforestation. The remaining half nudges the carbon dioxide concentration up each year and, with the other greenhouse gas emissions, is leading inexorably to global warming. It may be 250 years before a year's extra carbon dioxide is finally removed.

Photosynthesis

Plants absorb carbon dioxide from the atmosphere as a first step in the photosynthesis of carbohydrates. A proportion of that carbon dioxide contains carbon–14 and carbon–13, as well as the more common carbon–12. However, the stomata in the leaves give preferential treatment to carbon–12, which therefore slips through the cell wall. Since animals and decomposers such as fungi consume plants, they also finish up with more carbon–12 than you would find in the atmosphere.

Consequently we have the means here to detect how much fossil fuels and biomass we are burning. For instance, if we find that the atmosphere contains less carbon–13 than it did years, centuries or even millennia ago, this would suggest that it has been diluted with carbon–12, which must come from vegetation, whether living or fossilized. Also carbon–14, through changes in its relative proportions to carbon–12 and carbon–13, can indicate what proportion of carbon dioxide emissions can be attributed to fossil fuels rather than the burning of forests.

Photosynthesis across the globe actually mops up ten per cent of the atmosphere's carbon dioxide each year. If nothing happened to replace that carbon, then in matter of a decade no carbon dioxide would be left in the atmosphere and photosynthesis as we know it, would grind to a halt. In fact, the oceans contain fifty times more

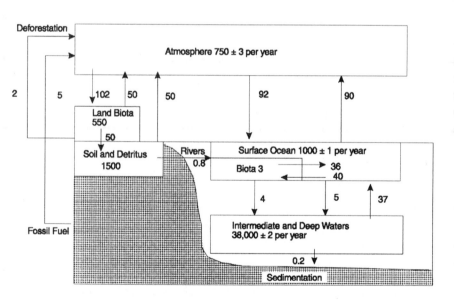

Figure 6. Global carbon reservoirs and fluxes. Units are gigatons. (Souce: IPCC, 1990).

carbon dioxide than the atmosphere, and as the concentration of carbon dioxide falls in the air, some escapes from the oceans. This *bubbling-out* is a slow process and certainly not as fast as the uptake of carbon dioxide by vegetation. Therefore, if photosynthesis continued at that pace, in less than one thousand years there would be too little carbon dioxide left in the atmosphere for photosynthesis to occur. If life could hang on, though much reduced in scope, perhaps we would observe a see-saw as carbon dioxide levels plummeted and then slowly rose again as the oceans released the gas into the atmosphere.

However, that is not what we see, and any such precipitous decline does not occur for the simple reason that organisms are not alike. If photosynthesis is the constructive phase of life, then respiration, in which the products of photosynthesis are decomposed, is the destructive counterpart. Respiration releases energy from the products of photosynthesis, and so we regain the water and carbon dioxide that plants forged into carbohydrates. An annual plant that germinates, grows, flowers, seeds and then dies within a year, will put back practically all the carbon dioxide that it had photosynthesized during its final decomposition by bacteria, fungi or even a grazing cow. A fraction of its productivity remains in the seeds, ensuring continuity into the future. The same is true for longer-living plants, such as trees living for hundreds if not thousands of years. A good proportion of the products of photosynthesis are consumed by the plant itself for maintenance. Total decay follows death and in the end the cycle of creation and then destruction is completed. Should a surplus exist, it is soon mopped up by opportunistic organisms, including ourselves.

Respiration

It is extraordinary how closely photosynthesis and respiration are balanced, which indicates that some kind of regulation is taking place. Charles Keeling, at the Mauna Loa laboratory, played a pivotal role in alerting us to the global dynamics of photosynthesis and respiration. Carbon dioxide is the key and Keeling found that rather than obtaining a straight line of carbon dioxide concentrations in the atmosphere during the months of the year, the line followed a regular, though oscillating path, with an upward pulse that reached its peak in May and its trough in October. In 1958, when Keeling and others first established the Mauna Loa laboratory, carbon dioxide levels had reached 315 ppmv. Forty years later, levels were up 50 ppmv. The annual pulse now raises the concentration by 7 ppmv in May, from its low in October. Keeling interprets this pulse as a demonstration of seasonal cycles of growth and decay, therefore of photosynthesis and respiration.

In fact, the amounts by which the pulse rises and falls each season are not exactly equal and the trend is ever upward. We suppose that the small, but detectable boost upwards is caused by industrial and agricultural emissions, particularly the burning of fossil fuels. Where the dynamics of carbon dioxide fluxes and flows have been followed elsewhere, they are very distinct from the data obtained at Mauna Loa. Such findings do not cast doubt on the quality of the Mauna Loa laboratory. On the contrary they illuminate a phenomenon that was not foreseen: first, that there was a pulse at all and second, that the amplitude of the seasonal pulse is greater as one travels north and measures carbon dioxide concentrations in the atmosphere above the boreal wastelands. What we are seeing is a veritable burst of photosynthesis that takes place when spring finally arrives. Above Alaska the pulse is at least twice that at Mauna Loa, and south of Mauna Loa the pulse tapers off to about 2 ppmv and nearly vanishes altogether below the Equator, and south to Antarctica. Clearly the larger land mass in the northern hemisphere is largely responsible for the size of the pulse.

The pulse was not always as large and when Keeling first encountered and measured it in the late 1950s and early 1960s he found it to be closer to six rather than seven parts per million, as it is now. That rise of more than sixteen per cent in forty years indicates a fairly dramatic change. Since both the ups and downs of the pulse are up, that must mean that both photosynthesis and respiration have increased. From experiments with plants grown under different temperatures and carbon dioxide concentrations, biologists have found that higher levels of carbon dioxide stimulate photosynthesis, but that soil organisms, such as bacteria, fungi, worms and termites, are also more active at higher temperatures. The two processes therefore more or less cancel each other out, which explains why in the Tropics, where seasonal changes in temperature are virtually absent, the carbon dioxide pulse is barely evident.

Obviously changes are taking place. Temperature increase across the surface of the planet has certainly been detected as has the rise in carbon dioxide. Seasonal changes of a sufficient size to be detected at Mauna Loa, are also taking place at high latitudes. Keeling has also found that the downward limb of the pulse has advanced by one week over the past 40 years, and he suggests that spring is now coming a little bit earlier to the high northern latitudes. The pulse itself is now nearly forty per cent higher in Alaska than when measurements were first taken in the 1960s. Since we know that about fifty per cent of our carbon emissions remain in the atmosphere, amounting to 1.8 ppmv per year, clearly, we have to look to ourselves as the prime suspect for such dramatic and recent changes.

Origins of Photosynthesis

Photosynthesis came about three and a half billion years ago, when blue-green bacteria, the *cyanobacteria*, produced the chlorophyll molecule, together with the chemistry for using light to split water so that the reactive components could then combine with carbon dioxide gas and generate carbohydrates. Oxygen was one by-product of this remarkable biochemical feat. As the bacteria colonized more and more of the Earth's surface increasing amounts of oxygen were released until the gas began to build up in the atmosphere. Initially, a considerable proportion of the oxygen reacted with exposed rock, oxidizing it and being removed from the atmosphere in the process. Then, photosynthesis gained the upper hand and, around two billion years ago, atmospheric oxygen concentrations started to rise to their current levels of around 21 per cent.

Photosynthesis was, and is, critical for life on the planet. It enabled a constant stream of energy to be tapped as well as the colonization not only of the surface of the ocean, but also of the land. James Lovelock's Gaia theory (see Chapter 8) suggests that without the all-pervasive colonization that has resulted from photosynthesis life would not be capable of regulating conditions on the planet's surface. Without such regulation, especially of surface temperature, the planet might no longer be the habitable place that it is.

Oxygen is potentially an extremely toxic and dangerous gas. Its ability to oxidize, especially when in the form of products such as hydroxyl (OH) and single oxygen atoms, means that virtually all chemicals in the body of an organism are at risk. On the other hand, the taming of that oxidizing potential has given organisms the power to extract energy from chemicals such as carbohydrates. Organisms with high metabolisms, such as ourselves, need both the oxygen in the atmosphere and chemicals with high-energy bonds. We call that

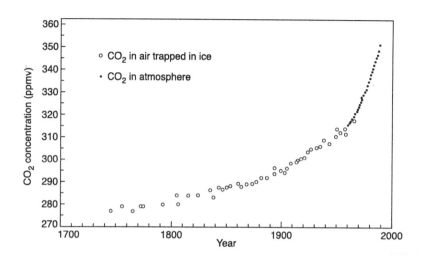

Figure 7. Atmospheric carbon dioxide increase in the past 250 years, as indicated by measurements on air trapped in ice from Siple Station, Antarctica, and by direct atmospheric measurements at Mauna Loa, Hawaii.

process *respiration*.

Without free oxygen this planet might have begun to lose its water. If that loss continued, perhaps our planet would suffer the same fate as Venus and Mars. It was James Lovelock who first asked the question why Earth's flanking planets did not have water when Earth did. He suggested that free oxygen released from photosynthesis interacted with free hydrogen to form water, so effectively blocking hydrogen's escape.

Hydrogen is released from volcanoes as water hits red-hot basalt. If the atmosphere were free of oxygen, there would be little to prevent the hydrogen wafting upwards and escaping into space. In an atmosphere that has free oxygen, the chances are that it will interact with hydrogen, and is therefore captured as water. If the oxygen was not there, generated by photosynthesizing organisms, then little by little the planet would lose its water and finish up as dry as its neighbours. As long as life exists and carries on photosynthesizing we shall have water.

Glaciers and Fossils

As evidence of solar activity goes hand-in-hand with historical records of climate, so too do measurements of greenhouse gases. We are fortunate that snow falling over frozen wastelands, such as the tops of high mountains and polar regions, carries minute bubbles of contemporary air with it that stay trapped, and therefore provide an extraordinary glimpse of the atmosphere as it probably was tens, if not hundreds of thousands of years ago. Analysis of air bubbles trapped in the glacier ice of Greenland and Antarctica not only confirms the measurements of carbon dioxide from Mauna Loa, but indicates that the pre-industrial levels of carbon dioxide during the eighteenth century were around 275 ppmv, and that over the past 160,000 years the levels of the gas never exceeded 300 ppmv until the mid twentieth century.

Oxygen Isotopes

Fossil ice tells us more than just the concentrations of gases in the atmosphere. By looking at the ratios of different oxygen isotopes in water scientists can get a good idea of the air temperature at a particular date in history. Oxygen–18, for instance, is 12.5 per cent heavier than oxygen–16, and most of the oxygen in the atmosphere is the lighter isotope. Because of its additional mass oxygen-18 confers ten per cent extra weight to water and therefore has a slight tendency to get left behind when water evaporates. The reverse is also true: that heavier water, with its atom of oxygen–18, will condense at a slightly faster rate than its lighter equivalent. Physicists have now worked out how the rates of evaporation and condensation are influenced by the temperature at which these processes occur. In fact, the concentration of oxygen-18 in the ice caps declines by roughly one part per thousand for every degree drop in the average surface temperature.

Isotopes and Ice Caps

The physics of looking back at past climate has become extraordinarily sophisticated and the isotopes of carbon and oxygen have proved particularly important. Today, climatologists believe they can measure how much water existed as ice at any moment during the past 160,000 years — the limit to how far back they can go being the historical length, and hence age, of the ice core. To do the calculation the researcher has to find evidence of the oxygen–18 content of sea water that actually corresponds with an ice core sample. However, unlike ice, water does not remain neatly segregated one year after another. This is where radiocarbon takes effect. In time, carbon links with oxygen to form carbon dioxide which then washes out of the atmosphere as carbonic acid and finds its way into the ocean. Some of that carbon, as bicarbonate, is incorporated into the calcareous shells of minute algae which, on dying, collect on the ocean floor and are stratified into layers of limestone deposits, providing researchers with the time sequence that they need. By looking at the fossilized calcareous shells in any particular layer, they can date the actual year when that layer was formed. Meanwhile, some of that same carbon dioxide will have oxygen-18 atoms and by comparing the ratio of these atoms with the ratio in ice deposited at that time, it can be determined how much water was in the oceans and how much was ice.

The Vostok Ice Core

The conclusion that carbon dioxide concentrations are rising at rates and to levels that are unprecedented over the past 250,000 years depends extensively on data derived from Antarctic ice cores. In the early 1980s Soviet scientists drilled 2000 m down into the ice sheet at their Vostok research station, and, despite the Cold War, collaborated with the United States airforce to transport the intact, frozen cores to France's glaciological laboratory at Grenoble, where they were analysed under the guidance of its director, Claude Lorius. The extremely low temperatures at Vostok — around -89°C — ensured that the record of ice formation was complete in that no melting was presumed to have occurred over that entire length of time. The Vostok core has provided some of the most concrete evidence of atmospheric changes that have taken place over two ice ages and interglacials. Scientists have carried out comparable work on ice core samples from Greenland, and the results from the two poles are in relative agreement regarding the timing and nature of swings from ice age to interglacial and back again.

The interesting finding is that the fluctuations of carbon dioxide at a particular time in the sequence closely match the rise and fall of temperature. In particular, the ice cores from Greenland indicate

TIME HORIZON		
	20 years	*100 years*	*500 years*
Carbon dioxide	1	1	1
Methane	63	21	9
Nitrous oxide	270	290	190
CFC-11	4500	3500	1500
CFC-12	7100	7300	4500
HCFC-22	4100	1500	510

Figure 8. Global warming potentials. The warming effect of an emission of 1 kg of each gas relative to that of carbon dioxide.

changes of around 50 ppmv in the concentration of carbon dioxide that seem to have taken place in less than one hundred years and are associated with abrupt temperature changes of as much as 5°C. Were the highs of carbon dioxide during interglacials a response to the high temperatures of the period, or were they the cause of the high temperatures? We can ask similar questions with regard to the low concentrations of carbon dioxide during ice ages. Feedback processes are certainly involved by which a change in one parameter, such as solar energy, causes a change in another parameter, such as greenhouse gas concentration; all of which results in the accentuation of the original parameter.

Tropical Glaciers

Glaciers offer an advantage over the polar ice caps in that they also exist in high mountain ranges in the Tropics, such as the Andes of South America, and therefore provide some indication of past climates in equatorial rather than polar regions. Three American scientists, Lonnie Thompson, Mary Davis and Ellen Mosley-Thompson have taken core samples from two glaciers in Peru, Quelccaya in the eastern Andes and Huascarán in the Cordillera Blanca, finding sharp fluctuations between wetter and drier regimes, which had a profound impact on local civilizations. When rainfall was high in coastal areas, civilizations thrived there; when rainfall gave way to drought, the civilizations collapsed. Conversely, highland civilizations thrived when the coast was in the throes of drought, and did less well when it was raining on the coast. In such marginal environments switches in climate appear to have direct consequences on human survival and therefore on cultures. More dramatic changes are now afoot and Lonnie Thompson and her colleagues have recently found disturbing evidence of global warming. For the first time since they took the ice core samples from Quelccaya in 1983, the glacier surface is now melting and water is finding its way vertically down the ice cap. Even though the glaciers are unlikely to disappear for some decades to come, 1500 year records of the past are being destroyed.

Global Warming Potentials

Aside from water vapour, carbon dioxide is the most important greenhouse gas in the atmosphere, with its concentration more than one hundred times greater than that of methane, and one thousand times greater than that of nitrous oxide. When water vapour is taken into consideration, CO_2 is responsible for approximately one quarter of the greenhouse effect. The current levels of CO_2 in the atmosphere are sufficient to absorb nearly all the radiation in the 15μm band of infra-red emanating from the Earth's surface. Consequently, more CO_2 in the atmosphere has little added effect around the midpoint of the absorption range. It is only at the edges of the range, where saturation has not yet occurred that additional CO_2 can still be effective as a greenhouse gas. Climatologists therefore take such effects into account when modelling the consequences of additional concentrations of greenhouse gases.

Each greenhouse gas makes its own particular contribution to global warming, which it continues to do until washed out of the atmosphere by rain, or more likely is broken down through chemical interactions, some of which may be catalyzed by sunlight. Sometimes those interactions lead to the production of other greenhouse gases — for instance, methane oxidizes to form carbon dioxide and water. Meanwhile, a gas, like a CFC, may be present in the atmosphere in very low quantities, but still have a significant effect. Climatologists therefore invoke the idea of *Global Warming Potentials* in which the impact of emitting 1kg of a gas over a stretch of time, such as one hundred years, is compared with that of carbon dioxide. The global warming potential therefore takes into account the disappearance of the gas from the atmosphere over time. Global warming potentials are likely to increase in the future as carbon dioxide builds up in the atmosphere. The increase occurs because of saturation effects. Thus, relative to carbon dioxide the effects of other greenhouse gases will proportionately increase.

Methane

After carbon dioxide, methane is the next most important greenhouse gas, and one of the keys to the climate of the past. Methane is generated by anaerobic bacteria that have a history going back virtually to the beginning of life on Earth. Methanogenic bacteria, with their need for oxygen-free places, such as bogs, swamps, irrigated rice paddies or the guts of animals, recall a time, more than two billion years ago, when the Earth's atmosphere lacked oxygen and methanogens happily exposed themselves to the atmosphere. Once photosynthesizers had evolved and oxygen began to build up in the atmosphere, the methanogens' intolerance of oxygen forced them to go into hiding.

Methanogens survive by pulling the oxygen atoms off carbon dioxide and then directing the oxygen to recombine with hydrogen or other hydrogen-rich chemicals, releasing energy in the process and fuelling their metabolism. They also convert buried organic matter into the gas methane. Methane, when in the atmosphere, tends to interact with oxidizing molecules, of which the most important is hydroxyl (OH). Hydroxyl derives primarily from the interaction between ozone and water, and by oxidizing methane, hydroxyl converts to water and in essence methane therefore removes oxygen from the atmosphere.

In what is one of the strange paradoxes of life on Earth, bacteria that shun oxygen actually play an important role on oxygen levels in the atmosphere. Without methanogens oxygen levels would begin rising to the point when soaking wet organic matter would be in grave danger of spontaneous combustion. Even the wettest forests in the world, like the Chocó rainforests lying off the Pacific Coast in Colombia and Ecuador, with their 12m of rain each year, would burn, without these ancient anaerobic bacteria.

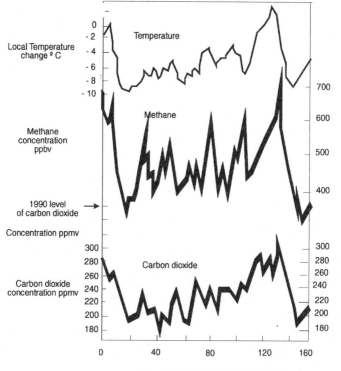

Figure 9. Changes in the estimated carbon dioxide and methane atmospheric levels over the past 160,000 years indicate a close correlation with changes in local temperature. (Source: IPCC, 1991).

Production of Methane

Methanogens are prolific and they emit several hundred million tonnes of methane a year. Natural wetlands generate approximately one fifth of the total, with the rest coming from rice paddies as well as gut fermentation in termites and vertebrates, such as ourselves. Like everything else we have interfered with this natural system. Forests have been cleared to make grazing land for cattle, and we have created irrigated swampland for growing rice.

Ruminants, such as cattle, with their multiple stomach chambers for digesting cellulose-rich substances like grass, cattle are prolific producers of methane, particularly when the vegetation is poor, as is generally the case with pasture grown on the soils of what was once tropical rainforest. Cattle, points out biologist Lynn Margulis, can be seen as mobile 40 gallon methane generators. In fact, physiological tests on potential candidates for manned space exploration indicate that even some people generate as much as 30 litres of methane a day.

Other sources of methane include escapes of natural gas during drilling and exploitation, as well as from coal mining, landfill sites, biomass burning, methane hydrate deposits and also oceans and freshwater. Climatologists ascribe sixty per cent of methane emissions — close to 300 million tonnes — to man-made sources. Currently some 60 million tonnes a year of methane are accumulating in the atmosphere.

Methane was present as a greenhouse gas in the pre-industrial atmosphere of 1750, but at the relatively low concentration of 0.8 ppmv, its contribution to surface warmth was relatively small. With an annual rate of increase of nearly one per cent, the gas has now doubled its concentration to 1.72 ppmv thus exceeding its highest ever concentration in the past 160,000 years. The methane sequence shows interesting parallels with carbon dioxide, in that it also follows the ups and downs of temperature shifts. Warmer spells are associated with methane concentrations of up to 0.7 ppmv and cold spells with concentrations that fall down as low as 0.3 ppmv.

During the last ice age most of the methane would have been generated in the Tropics, since the marshy areas of the higher latitudes were covered in a thick sheet of ice. Some of the best evidence for methane concentrations at that time comes from Jerome Chappellaz, of the Grenoble laboratory of Glaciology and Geophysics. His investigation of the Greenland ice core reveals a dramatic thirty per cent drop in atmospheric methane during the last ice age. He believes that the fall in methane, combined with the evidence of a marked increase in atmospheric dust levels, are strong indications that the Tropics actually dried out during those periods.

On the other hand, the rise of sea levels associated with global warming could lead both to the sea encroaching on low-lying land as well as to swamping as the waters of rivers accumulate in estuaries. The waterlogging of new land areas would undoubtedly favour methane-generating bacteria.

Taking its average life span in the atmosphere of eleven years into account, molecule per molecule methane is seven and a half times more effective as a greenhouse gas than carbon dioxide, which, on a weight to weight basis, has a global warming potential of twenty-one over a hundred year period. However, the amounts of the two gases emitted into the atmosphere as a result of human activities are very different, and whereas we currently emit six per cent of all (6 GtC of every 100 GtC) carbon dioxide emitted by living matter each year, we are responsible for sixty per cent of total methane releases. The net result of applying methane's global warming potential to the estimate of the total emissions for which we are responsible suggests that methane's overall contribution to global warming is one quarter that of carbon dioxide.

Potential Methane Releases

Considerable quantities of methane are trapped under permafrost in boreal regions close to the Arctic Circle. Rising temperatures, resulting in the melting of permafrost, could bring about the release of as much as 450 GtC in the form of carbon dioxide and methane. Such a release would be a powerful, self-reinforcing feedback, since it would incur more temperature rise and permafrost melting and thus further releases. We already have considerable cause for concern. Siberia, much of which is covered in permafrost, is warming faster than almost anywhere else on the planet.

According to the US Geographical Survey, 10,000 billion tonnes of methane are currently trapped under pressure in crystal structures — methane hydrates — on the edges of continental shelves, making them the Earth's largest fossil fuel reservoir. Should the temperature in the surrounding water or sediment be increased to the point where methane hydrate becomes unstable, methane gas is released overnight. Hence, where water is relatively shallow and thus easier to heat, as in the Arctic (which is already warming at a rate two to three times the global average), tens if not hundreds of billions of tonnes of methane could be released.

Methanogens and Hydroxyl Radicals

The life span of gases in the atmosphere such as methane depend on chemical reactions involving oxidizing substances and energy from the Sun. The hydroxyl radical is generated when ultraviolet light splits ozone into oxygen (O_2) and an oxygen atom that immediately reacts with water. For every split ozone molecule two hydroxyl radicals are created. Much of this generation takes place in tropical skies over low latitudes simply because the ultraviolet of the appropriate wavelength is able to penetrate the low ozone levels above equatorial regions. The quantities of hydroxyl produced by photochemical reactions are extremely small, yet they are sufficient to account for the oxidation of methane, carbon monoxide (CO), nitrous oxide, hydrogen sulphide as well as traces of organic gases such as methylchloroform, which have got into the atmosphere through industrial emissions. The hydroxyl radical therefore plays a seminal role in cleansing the atmosphere of substances that might otherwise accumulate.

The outcome of different reactions involving hydroxyl depends largely on the concentrations of nitrogen oxides. When these oxides are present in ample quantities, the oxidation of each molecule of methane to carbon dioxide follows a number of steps that finally lead to nearly four molecules of ozone being generated and half an hydroxyl radical. On the other hand, if the concentration of the nitrogen oxides is low, the oxidation of each methane molecule leads to a net loss of more than three hydroxyl radicals and nearly two ozone molecules. Current research suggests that the nitrogen oxide rich environment and its counterpart, the nitrogen oxide poor environment, are of similar size. If so, any increase in methane in the atmosphere is likely to lead to a net loss in hydroxyl radicals; furthermore, such a loss will gradually accelerate as methane accumulates in the atmosphere.

As a result of atmospheric pollution from industry, we are also increasingly likely to see the removal of methane by interaction with hydroxyl taking place outside the Tropics and over northern mid-latitudes instead.

Ozone Depletion

Another consequence of the gradual increase in methane in the atmosphere is that greater amounts will escape oxidation in the lower atmosphere and gradually pass into the stratosphere. Oxidation of methane in the stratosphere leads to the production of water vapour which, apart from being implicated in chemical transformations that lead to the destruction of ozone, has the effect, as a greenhouse gas, of raising temperatures. But, while methane and water vapour are increasing in the stratosphere, ozone is being depleted, and ozone is a greenhouse gas. Scientists are still debating whether the net result will be stratospheric warming or cooling.

Chapter Four:
Winds, Volcanoes and El Niño

British Weather

The abrupt differences between one regime and another explain why the weather in Britain is such a perennial topic of conversation. The polar maritime north-westerly brings cool, often gale-force winds from northern Canada and the Arctic Ocean; the Arctic maritime originates from the North and, if it prevails, can bring bitter cold; the polar continental, with its blasts of wind from Siberia, is often associated with heavy snow falls; while the westerlies stem from the tropical maritime and bear warm, wet weather; finally, the tropical continental strikes Britain from the South, which, during summer, is

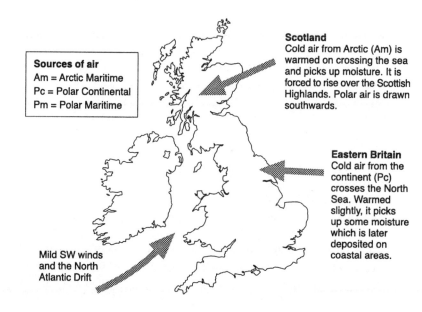

Sources of air
Am = Arctic Maritime
Pc = Polar Continental
Pm = Polar Maritime

Scotland
Cold air from Arctic (Am) is warmed on crossing the sea and picks up moisture. It is forced to rise over the Scottish Highlands. Polar air is drawn southwards.

Eastern Britain
Cold air from the continent (Pc) crosses the North Sea. Warmed slightly, it picks up some moisture which is later deposited on coastal areas.

Mild SW winds and the North Atlantic Drift

Figure 10. British wind patterns.

likely to expose the country to a heatwave.

Air masses tend to stick to their own and when one encounters the other, the warmer rides over the cooler, either forcing its way over or being pushed up, depending on which of the two air masses is advancing. Should the warm air predominate, then that front is known as a *warm front*; if cold air forces the encounter, then the front is a cold one. Whichever it is, the consequences tend to be the same: the warmer air cools as it passes over the other and clouds form as water vapour condenses, and rain may follow. In general, the polar front arises when warm, moist tropical maritime air meets cold polar maritime air from the Arctic Ocean. As the warm air rises over the cold, denser, northerly air it leaves behind a region of low pressure, known as a *depression*.

Air Circulation

Towards the end of the seventeenth century the English mathematician and astronomer, Edmund Halley realized that the trade winds and monsoons were the result of the Sun heating the atmosphere differentially. These winds are absolutely crucial for distributing the Sun's energy, and making the polar regions warmer than they would be were there no air to transport the heat from one part of the planet to another. The Equator receives nearly two and a half times more energy than the poles over the course of the year. If that energy were not distributed, the Equator would be 14°C warmer and the North Pole 25°C colder than at present. The atmosphere, with considerable help from the oceans, effects the transfer of energy from the Equator to the poles, thereby generating the world's climate in the long term and its weather in the short term.

John Hadley, a contemporary of Halley, took his ideas a little further in devising the notion of a giant convection cell in each hemisphere that began at the Equator and travelled in a loop towards the poles and then back down.

As a result of the spinning of the Earth the nineteenth century engineer and mathematician, Gustave Coriolis, realized that were any winds travelling north would be deflected by the spin and return to the Equator, more or less following the direction of the trade winds. Friction between the Earth's surface and the air immediately above it tends to counteract the Coriolis force, but it is a weaker force that falls off with altitude.

Both Halley and Hadley were mistaken in thinking of the general

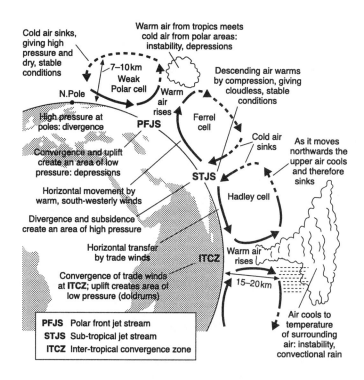

Figure 11. Atmospheric circulation in the northern hemisphere.

air circulation in each hemisphere as one giant convection cell. To account for the tropical trade winds in the Atlantic and Pacific Oceans, for the south and north-westerlies in each of the two hemispheres and the easterlies running from the Polar regions to the mid-latitudes, William Ferrel suggested in 1856 that each hemisphere had to have three convective cells. Carl-Gustaf Arvid Rossby refined that idea in 1941 and added his notion of high altitude jet streams of air. As confirmed by satellite, the meandering Rossby waves between high and low pressure zones move whole weather patterns into different latitudes and countries. Their position and behaviour can tell meteorologists a lot about the weather.

The convection cell which circulates in a clockwise direction between the Equator and 30° N has been given Hadley's name, as has its mirror image in the southern hemisphere. Air rises at the Intertropical Convergent Zone (ITCZ) and flows in a north-easterly direction in the northern hemisphere and a south-easterly direction in the southern hemisphere, in both instances creating a zone of low

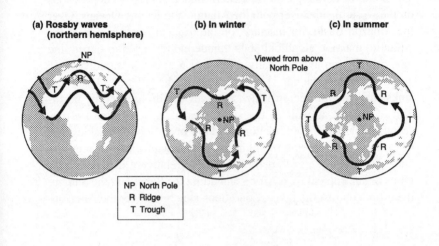

Figure 12. Rossby waves and jet streams in the northern hemisphere.

pressure which draws in air that is supplied by air returning from the Tropics, generating the trade winds. Laden with vapour, the merged trade winds are pulled up into the sky to form a tropical belt of clouds that in turn irrigate the rainforests of the equatorial zone. Meanwhile, the movement of the overhead Sun over the course of the year causes the ITCZ to shift from one side of the Equator to the other; being most southerly, over the Tropic of Capricorn, in December, when the southern summer reaches its apogee, and most northerly in June, over the Tropic of Cancer, at the height of summer in the northern hemisphere. The solstice shift between the two hemispheres first strengthens the trade wind and then weakens it.

The *Ferrel cell* is the second circulation cell in the system: it circulates in the opposite direction to the *Hadley cell*, with the high pressure downward limb responsible for the gusty warm south-westerlies in the northern hemisphere, and the 'Roaring Forties' in the southern hemisphere. The Ferrel cell's upward limb generates a low pressure zone at a latitude of 60°. Finally the *Polar cell*, moving in a clockwise motion, carries warmer air upwards towards the pole and back down, thus generating the north and south-easterlies of the respective hemispheres.

The Hadley cell is therefore responsible for dissipating the high energies received over the Equator and we find typically high pressure zones developing over the eastern side, and massive depressions on the western side, over both the Atlantic and Pacific Oceans. Over the Amazon Basin, for instance, on the tropical western side of the Atlantic, massive cumulonimbus thunderclouds develop during the day, bringing an average daily rainfall of 6 mm. The rapidly rising air of the thunderclouds is pulled out over the Atlantic Ocean and sinks over a wide region which includes the drought-prone northeast of Brazil, as well as the subtropics of west Africa. The descending part of the circulation, tends to bring cloudless skies and dry weather; the average rainfall associated with the high pressure air is less than one millimetre a day. The relatively dry, desert regions of subtropical west Africa, including the Sahel and the Sahara, therefore contrast the humid conditions generated over the Amazon.

Blown off Course

During World War I, pilots trying to navigate Zeppelins, found themselves blown off course when they ventured high in the troposphere over the middle latitudes. World War II pilots, heading north or south at altitudes above 8 km over similar latitudes, also found it hard to maintain a course because of strong crosswinds. At the same time, they found that eastward-bound flights took far less time than flights in the opposite direction. Rossby soon realized that a narrow band of strong westerly winds, with speeds of 200 kph and more, was circumnavigating the planet in the middle latitudes. Later investigation showed this circumpolar vortex to follow a meandering path as it travelled around the globe. Rossby's account of the wave-like motion of the air mass, with its ridges and troughs, came to be known as *Rossby waves*. The ridges in the system represent warm air carried north from the Tropics, and the troughs represent cold air that has replaced the warmer air. The faster bands of air, some travelling at speeds of more than 230 kph, within the circumpolar system, are the *jet streams* that the pilots encountered.

The vortex is , in fact, extraordinarily deep, carrying a mass of air that may range from 2 km above the surface to as much as 20 km above the Earth. Much of the momentum of the atmosphere is caught up in this single massive air system within each hemisphere.

Jet Streams

As if wedging through the division between adjacent convective cells, jet streams can spread the ash and sulphurous material from an erupting volcano around the Earth within a week, thereby potentially affecting the weather on a planetary scale. The polar front jet stream lies between 40° and 60° in both hemispheres, thus forcing its way between the rising arms of the Ferrel and Polar cells, between the warm humid air of the Tropics and the cold dry air of the poles. The exact position of the jet stream makes a considerable difference to the weather and when, in the northern hemisphere, the stream moves south it carries cold air with it which sinks in a clockwise direction, giving rise to the dry, stable conditions that we associate with high pressure anticyclones. When the same jet stream shifts northwards it is now warm and rises in an anticlockwise direction, bringing with it the strong winds and heavy rainfall associated with depressions and areas of low pressure.

The usual path of the polar front jet stream over Britain is from the south-west towards the north-east, which accounts for much of Britain's more miserable weather. This is occasionally alleviated by anticyclones that block the jet stream forcing it southwards, so that while Britain bathes in hot, dry weather, such as the summers of 1976, 1989 and 1995, regions further to the south are drenched in unseasonal rain.

El Niño

The cycle of one El Niño every few years certainly fitted the pattern of a system that released its energy in a single event. However, the cycle has changed dramatically over the past 30 years, with four consecutive El Niños since 1990, and more El Niños than 'normal' since 1976. Kevin Trenberth and Timothy Hoar, of the National Centre for Atmospheric Research in Boulder, Colorado, believe the last El Niño sequence, which ended in June 1995 only to restart with a vengeance in 1997, has been the longest for about 2000 years.

Is the succession of El Niños an indication that global warming is truly underway? If so, then the climatic consequences could be drastic, especially in drought-prone Africa and in countries such as India which depend on the monsoon. It may well be that a succession of

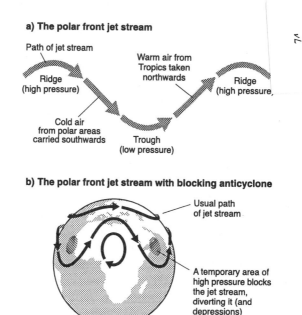

a) The polar front jet stream

Path of jet stream

Warm air from Tropics taken northwards

Ridge (high pressure)

Ridge (high pressure)

Cold air from polar areas carried southwards

Trough (low pressure)

b) The polar front jet stream with blocking anticyclone

Usual path of jet stream

A temporary area of high pressure blocks the jet stream, diverting it (and depressions) southwards

Figure 13. The polar front jet stream in the northern hemisphere.

El Niños does indicate a significant shift in the energy balance of the planet in general, and the Pacific Ocean in particular. Paradoxically, however, the weather patterns that prevailed during abrupt cooling of the northern hemisphere in the last ice age, share many features with the El Niño of 1982–83 — dry weather over the Tropics and heavy precipitation over the mid-latitudes.

The latest El Niño involved massive releases of energy from the Pacific Ocean. The first signs in 1997 were a sudden warming of the tropical waters off the coast of Peru by as much as 6°C, combined with violent downpours and landslides in what is one of the driest deserts in the world — the Atacama.

From early July to August 1997, heavy rains over eastern Europe filled rivers to bursting point and sent waters swirling through towns, villages and farms. Bridges were destroyed, communications disrupted and by the fifth day at least nineteen people had died, with tens of thousands made homeless. In September, on the other side of the world in East Kalimantan, southern Borneo, a catastrophe of a different kind occurred. Tens of thousands of Indonesian farmers set felled trees alight to make clearings for their gardens, just as they did every year. However, in 1997, the fires raged out of control. The usual summer rains did not materialize and for weeks clouds of acrid smoke drifted across Malaysia, Thailand and the Indonesian islands of Java and Sumatra, blocking out the Sun and making life a misery for millions of people.

There were devastating hurricanes on the Mexican west coast in the autumn, followed by ferocious typhoons, such as Typhoon Linda that killed hundreds if not thousands in Thailand, South Vietnam and Cambodia early in November. That same week saw a 150 mph tornado in Florida, and landslides in the Azores brought on by torrential rain, were followed by violent storms over Spain and Portugal, and unprecedented December floods. Even the western British Isles suffered storm-force winds that continued into January 1998.

Lack of rain was widespread over South-East Asia in 1997 and by the end of October, Papua New Guinea was in the throes of a terrible drought that threatened much of the population with starvation. The Amazon, also showed signs of drying up and the early part of 1998 proved to be the driest period for decades, putting whole sections of rainforest under considerable stress. By May 1998, massive fires, started by cattle ranchers, had broken out in Roraima, Brazil's most northerly, Amazonian state, and in Mexico, in the state of Oaxaca, fires covering several hundred thousand hectares were also raging out of control.

It is no coincidence that the amount of carbon dioxide accumulating in the atmosphere in 1998, as measured at Mauna Loa, Hawaii, jumped upwards by one seventh compared to the previous peak.

The Harbinger of Worse to Come

El Niño hits hardest where we have already begun to degrade the environment. The horrendous damage and loss of life in Honduras and Nicaragua caused by mudslides resulting from Hurricane Mitch, were largely the result of deforestation, which left soils exposed and vulnerable to sheet erosion and massive landslides. Having lost 34 per cent of its pine and deciduous forest through logging between 1964 and 1990, Honduras is still continuing to destroy its native upland forests at the rate of 80,000 hectares per year. The situation is even worse in Nicaragua where 150,000 hectares of forest are destroyed each year as a result of commercial timber extraction, the advancing agricultural frontier and slash-and-burn farming. The country has lost nearly sixty per cent of its forest cover in the last fifty years.

Trade Winds & Pacific Currents

El Niño is not unexpected but not all El Niños have been as devastating as the current one. We have experienced two El Niños over the past 15 years that appear to be stronger than any recorded over the past five centuries, and the suspicion is bound to arise that exceptional El Niños must be a consequence of global warming. If so, they are a salutary warning that climate change may not take place gradually and benignly, but through abrupt transitions from one state to another. If El Niños becomes the norm they will undoubtedly play havoc with our expectations of climate and weather patterns. Agriculture could be jeopardized; where before we had plenty of rain we could experience arid conditions and, vice versa.

In the 1920s Sir Gilbert Walker came up with some of the first meteorological descriptions of El Niño and the way in which the surface currents in the tropical Pacific switched as a result of the trade winds weakening. Like the Peruvian fishermen before him he observed how the failure of the trade winds resulted in much warmer water — as much as 5° or 6°C warmer — transforming the bright, dry weather from cool, descending air, to heavy rain. Just what causes the system to alter remains something of a mystery.

Causes of El Niño

As a result of Walker's observations, the switch of currents that causes El Niño is called ENSO —El Niño Southern Oscillation — and the particular circulation of the air masses associated with the Intertropical Convergent Zone (ITCZ) is known as the Walker circulation. In general, strong ENSO events affect weather and rainfall patterns over a large area of the globe; from monsoon rains that fall over the Indian subcontinent during the summer, which are essential for good harvests, to precipitation over the Amazon Basin which maintains the tropical rainforest. During an El Niño, the low pressure region of the Walker circulation shifts westwards and instead of rising over the western Amazon as it pushes up against the Andes, it rises over the Pacific Ocean. Meanwhile, the high pressure branch, with cooler air, covers much of the Amazon and reaches across to the west coast of Africa.

The southern Pacific Ocean mops up energy in the form of heat, until a point is reached when the system overloads and deposits the energy in one dramatic moment. As they warm, the surface waters of the central Pacific can move hundreds, if not thousands, of miles eastwards. Although the mechanism is not clear, the extra heat going into the oceans because of global warming appears to have tipped the balance towards more frequent El Niños, as well as more severe ones. El Niños therefore reflect an instability in the climate system and thus signal that climate is changing dramatically.

The amount of moisture in the climate system is a good indicator that the Earth is heating and, according to scientists at the US Global Change Research Program, atmospheric moisture has risen by five per cent per decade since 1973 over the United States, and by ten per cent over temperate regions of the northern hemisphere over the past century. More water vapour in the atmosphere means 'a significant increase in the energy available to drive storms and associated weather fronts,' such as El Niño.

Sea surface temperatures also indicate whether the energy content of the oceans and atmosphere is changing. A warmer ocean increases the build-up of moisture in the atmosphere and the subsequent release of latent heat when the moisture condenses into clouds and rain. The Pacific Ocean has warmed over the past century, with the temperature of waters in the central ocean having increased by at least 0.5°C.

The Volcanic Catalyst

Global warming on its own, however, may not be enough to cause an El Niño. One idea that has been gaining ground is that some of that initial energy comes from volcanic activity in the ocean floor. Oceanographers have discovered large flows of magma from the mid-ocean Pacific Ridge that over a period of approximately five years could release as much as ten per cent of the heat normally associated with changes in the sea surface, which accompany El Niños.

Even more extraordinary is the link with volcanic eruptions that send debris and gases into the atmosphere. Major volcanic eruptions, such as El Chichon in Mexico in 1982 and Pinatubo in the Philippines in 1991, sent so much debris into the atmosphere that as much as ten per cent of sunlight could no longer reach the Earth's surface over the northern Tropics.

According to Paul Handler and Karen Andsager, of the University of Illinois, the subsequent cooling led to substantial shifts in the amount of air building up over the Eurasian continent during the winter months. As a result, the air mass that normally feeds the trade winds was much weaker and conditions were set for an impending El Niño. On the basis of such a scenario, volcanoes are most effective in bringing about an El Niño when they erupt over the low latitudes of the northern hemisphere, as El Chichon did in 1982. The theory certainly held for 1997, which saw the eruption of Soufrière on the Caribbean Island of Montserrat.

On average, one major volcanic event occurs every decade and the ensuing fall in radiation leads to a drop in temperature that can be picked up as *frost rings* in trees in exposed locations. Usually cooling manifests itself a couple of months after the eruption. In general, because of the greater proportion of land to ocean, the northern hemisphere shows greater cooling than the southern hemisphere, where the heat stored in the sea acts as a buffer against temperature change.

Deforestation and El Niño

Whereas volcanoes and the 4-7 year El Niño cycle are all natural phenomena, we have now introduced the massive and continuing destruction of tropical forests. Tropical forests are responsible for prodigious releases of energy in the form of water vapour. This is the equivalent of the energy that would be released by detonating 6 million atomic bombs every day over the Amazon Basin. This energy is then transferred by global circulations from the Equator up into the higher latitudes, and is crucial for the movement of air masses.

The tropical forests of the world, including those of the Amazon, of Central Africa and of Indonesia, lie at the points along the Equator where tropical thunderstorms develop, but are not fortuitous recipients of rain; they actually generate rain. Firstly, they pump water vapour into the atmosphere through transpiration, and secondly, they release volatile hydrocarbons, such as isoprene, which act as *cloud condensation nuclei* (CCNs). Consequently, as much as three-quarters of all rain that falls over the rainforest in the Tropics is returned to the atmosphere by means of evapotranspiration.

Therefore, heated air masses rise above the forests and become rivers of air which cross the Pacific from west to east. They then cool down and descend where the waters are coolest, thus feeding the trade winds. Rainforests therefore act as thermal machines and, above all, as regulators of atmospheric and oceanic systems which control the climate.

Even though the El Niño Southern Oscillation has existed as a phenomenon for longer than history, the wholesale destruction of tropical forests over the past forty years has seriously jeopardized the efficiency with which energy is transferred from the Equator to the higher latitudes. Over the past 15 years, climatologist, Ann Henderson-Sellers, has attempted to model rainforests into the climate system. Her models indicate that rainforest destruction is having a significant impact on the jet streams that wedge their way between various atmospheric circulation cells. By shifting the air masses of the major circulation systems south and north, east and west, the jet streams have a profound effect on regional climate. Forest destruction in the Tropics is therefore changing climate and sending weather systems spiralling off in new and unpredictable directions.

Given the crucial role of tropical forests in re-distributing the energy that falls over the Equator, and man's continued destruction of them, we could be in for a spate of destructive switches between El Niños and La Niñas; these are the other side of El Niños, when the trade winds strengthen and the cold Humboldt current begins upwelling along the South American coast. A series of powerful El Niños would play havoc with tropical agriculture, with vast areas of the Tropics becoming desert as a result of successive drying out.

Extra heat trapped in the atmosphere is likely to alter where and how clouds form and how extensive they are. Conceivably clouds could block the build-up of high pressure over the Eurasian continent in the winter, and over the east Pacific ocean in the summer. El Niños could therefore follow on from year to year, and may be an indicator of a climate system that is operating close to an extreme of either warming or cooling. In either case the result is the blocking of deep water currents.

If the Walker and Hadley circulations did not transport humid columns of air towards the higher latitudes, temperatures over the Equator would soar during the daytime and plummet at night. The transport of atmospheric water from the Amazon Basin is of considerable climatic significance in the transfer of the Sun's energy; dry air is not only a poor conductor of heat, but carries far less energy in a given volume compared with water vapour. The energy required for enough vapour to condensate and fall as 2 cm of rain is sufficient to warm the entire tropospheric column by as much as 6°C. Sunshine over that same column of air would at best heat it up by 2°C, most of which would then be lost through radiative cooling.

This energy transport system depends critically on the rainforest, which pumps enormous quantities of water vapour into the atmosphere. It is now disturbingly apparent that the destruction of the world's tropical rainforests, and of the Amazon in particular, will have a devastating effect on climate.

Carbon Emissions from Deforestation

The tropical forests of Central and South America are unique in their capacity to grow, even when seemingly mature. Oliver Phillips and his colleagues report in *Science* that they have measured as much as one tonne per hectare per year of growth in these intact forests. Consequently, John Grace from the University of Edinburgh, estimates that if all the forests of the Brazilian Amazon, covering 360 million hectares, put on biomass in that way, the Amazon in Brazil alone would be an annual *sink* for up to 0.56 billion tonnes of carbon.

During 1998, 9 million hectares of tropical forest were destroyed by fire worldwide. On the basis that a hectare of tropical rainforest contains between 100 and 250 tonnes of carbon in its biomass — of which three quarters burns or decomposes — carbon emissions totalled between 1 and 2 Gt from that source alone, which is equivalent to one third of the emissions from fossil fuel burning across the world. When areas are cleared of trees the surrounding forest suffers die-back and disintegration. Carbon emissions from areas of the Amazon that have been cleared are likely to be at least seven per cent higher than previously thought, because that die-back is the same as felling one million more hectares than are actually cut down.

The Amazon as Heat Pump

Whilst the impact of tropical forest destruction on the uptake of carbon has now been modelled, climatologists have rarely taken into account an even more important and devastating consequence of this destruction: the process by which heat over the Tropics is carried away in massive rain clouds and distributed by air circulation towards the cooler, higher latitude regions. We are now discovering that without intact forest, the amount of solar energy that can be carried away towards the higher latitudes is reduced by a fifth or more. This cut alone would be sufficient to cause significant cooling over temperate zone countries such as Britain. Combined with the seizing up of the Gulf Stream the loss in heat transfer would be a devastating blow to the climate of Northern Europe and Scandinavia (see

Chapter 6). All tropical forests contribute to the process of energy transfer but, by virtue of its size — 5M km² in total — the Amazon is by far the most important of all tropical rainforests.

The incoming trade winds across the Atlantic pick up considerable moisture by the time they arrive at the Brazilian coast. When, for instance, the air warms from 25° to 30°C, the saturation vapour pressure increases by more than one third, which translates as seven grammes more water vapour being carried by each kilogramme of air. Warmer air therefore tends to result in more rainfall and more heat being released when the vapour condenses into water. In the mid-latitudes, a warming from 10°C to 15°C results in no more than half the change experienced at the higher temperatures.

According to the Brazilian physicist, Eneas Salati, between 50 and 75 per cent of all water falling as rain over the Amazon is evaporated and transpired back into the atmosphere. From here, it falls again as the winds of Walker circulation deposit it up to 7 times across the entire 4,000 km expanse of the Amazon Basin. Without such a mechanism, the air mass would become drier and devoid of water as it moves westwards. Yet the north-west region of the Basin receives twice as much rain on average than the central region, when the air circulation encounters the high ridge of the Andes. Some parts of the western Amazon have as much as 8 m of rain a year — nearly four times the average rainfall over the Amazon Basin.

To keep the air charged with vapour requires prodigious quantities of energy, all of which must come from the Sun. The energy carried away as latent heat each year from the Amazon amounts to more than 500 TW (TW = one million million watts) and therefore 40 times the total energy humans use worldwide for industrial and agricultural activities.

The average amount of the Sun's energy received at the Earth's surface is 240 Wm². Although in principle the Equator receives more than double the energy received annually by the poles, the high humidity and cloudiness over the Amazon Basin reduces the amount received at the top of the tree canopy. For about half the time during the day and in any one place the Basin is covered in clouds. Consequently the average amount of energy received annually at the top of the canopy is 181 Wm², about 25 per cent less than the global average. As much as three-quarters of the total energy available to the forest results in *evapotranspiration* — the process by which water evaporates from the surface and is actively pumped into the atmosphere through the stomata of leaves, which leaves one quarter of the available energy for heating the air.

In the rainforest of central Amazonia transpiration accounts for sixty per cent of the air's humidity, and evaporation for the remaining forty per cent. When the forest is intact virtually no evaporation occurs from the soil, but rather from the stems and leaves of the vegetation. Hot, humid air over the rainforest rises rapidly and develops into cumulonimbus clouds that irrigate areas further downwind and release the energy bound up as 'latent heat' back into the atmosphere. There the released energy drives the great air masses across the Amazon Basin until they reach the Andes. The flow then splits into three branches. The central part crosses the mountains into the Pacific and continues west along the Equator, following the convergence of the warm northern sea current; the southern stream is deflected by the Andes and passes over Patagonia via the Brazilian savannah, while the northern stream, crosses the Caribbean, touches the eastern seaboard of the US and goes over the Atlantic towards northern Europe.

The Rainforest and Rossby Waves

The polar front jet stream, the most northerly branch of the Rossby waves, is the most powerful of all the jet streams. It drives between the air masses that form in the polar region and those which form between the Tropics and the temperate zone. How far south the jet stream pushes makes all the difference to the weather. If the Hadley circulation becomes weaker because it has gathered less energy through deforestation, then the jet stream will have more power to force the entire weather system of the northern hemisphere down towards the Equator.

The Drying out of the Amazon Basin

If a significant proportion of the forest is destroyed the system of heat transfer will begin to collapse. J. Lean and P.R Rowntree of the UK Met Office find from their improved models that deforestation over the Brazilian Amazon could lead to as much as a 65 per cent reduction in rainfall over the Colombian Andes during certain seasons. They also corroborate Salati's contention that the regrowth of forest within areas that have been cleared, as well as the survival of forest in outlying areas, are likely to be threatened by an extended dry season, combined with less rainfall at other times of the year.

Deforestation has a major and immediate impact on the distribution of water. A 350 hectare tea plantation in tropical Africa showed a twofold increase in moderate flooding and a fourfold increase in more serious flooding compared with the nearby natural forest. The Brazilian climatologist, Luiz Carlos Molion points out that the Amazon forest canopy intercepts about fifteen per cent of the rainfall and that its destruction and removal would lead to as much as 4000 cubic metres (tonnes) per hectare per year of rain hitting the ground. Much of that water would run directly into the rivers, rather than being retained and maintaining soil moisture. The net result would be *sandification* whereby heavy drops of rain cause the selective erosion of finer clay particles, leaving increasingly coarse sand behind. With time, the remaining soil loses virtually all its water-retaining properties and the forest is unable to regenerate itself. Soil under intact forest absorbs 10 times more water compared with nearby areas that have had pasture for five years. Outside the forest and away from its soil-protecting attributes, erosion increases a thousandfold.

We have no idea just what proportion of forest must be left for the system to be self-maintaining. It may be three-quarters; perhaps even less: if so, with twenty per cent already gone, we are close to the limits.

Accelerating Destruction

Current estimates indicate that as much as 17 million hectares of tropical forests are being destroyed each year, when destruction from charcoal manufacture for pig-iron production is also taken into account. Aside from charcoal production, more than fifty million hectares of the Brazilian Amazon have gone in a matter of decades — a loss the size of France and one sixth of the total Brazilian rainforest.

The principal impetus for the destruction of the Amazon came with the promotion of cash crops such as soya for cattle feed in the country's southern states, with the result that hundreds of thousands of peasants were driven from their lands to seek their livelihood in the Amazon rainforest. There, because of the impossibility of sustainable agriculture better suited to subtropical and temperate soils, these peasants themselves became a destructive force. Once the Amazon rainforest in Brazil is wholly destroyed, the world will discover too late that it has pulled down one of the most important components of a stable, global climate.

Clouds and Forests

Thirty years ago we would never have imagined that life on Earth played a vital and important role in the generation of clouds. We now know that trees emit a range of substances, such as *isoprenes*, that act as the focal point from where water vapour condenses as droplets. Such cloud condensation nuclei (CCNs), are a necessary part of cloud formation and if they are not present, then clouds will not form. Forests, not only pump vapour into the air but also produce the nuclei around which that vapour aggregates into clouds.

The forest is therefore largely responsible for the recycling of water over land as well as maintaining the *hydrological cycle*. In addition, a tropical rainforest distributes solar energy to other parts of the planet and, in so doing, makes the planet a more habitable place.

Clouds are very much part of climate. If open water, such as the ocean, has a relatively dark colour that absorbs light and heat, then clouds above it tend to reflect light and heat, thus cooling the surface beneath them. But what generates clouds over the oceans? Where are the ocean's CCNs? In general, the atmosphere above the oceans is clean and CCNs are far and few between. Sea spray carries some sea salt sulphate into the air which makes excellent CCNs. However, such spray yields no more than ten CCN per cubic centimetre — not enough to generate the thick bank of marine stratus clouds that are found above the seas.

Dicovering the Oceans' CCNs

Nearly forty years ago, James Lovelock invented a device that in more than thirty years has not been superceded — the electron capture device. In the late 1960s, Lovelock was on holiday in western Ireland, testing his invention by showing that he could detect very low concentrations of industrial pollutants in the air. At that time, CFCs were just beginning to be used as refrigerants in fridges, freezers, air-conditioners and as aerosols in spray cans. The amounts used worldwide were very low but even though atmospheric concentrations amounted to only a few parts per billion, Lovelock's apparatus was still sensitive enough to pick up the presence of CFCs in the clean atmosphere over rural Ireland.

Out of curiosity he set out to measure the level of pollution on hazy days to compare with clear days. He surmised that the CFCs

would be excellent indicators of the extent to which man-made substances might spread throughout the atmosphere. He discovered that
on clear days the air contained fifty parts per trillion of CFCs, and
on hazy days carried three times as much. The haze was therefore of
industrial origin and had blown in from southern Europe. Moreover,
the amount of CFCs in the atmosphere suggested that they were not
being broken down.

This discovery of CFC contamination proved to be seminal in
awakening concern that CFCs could destroy ozone in the stratosphere. In 1971, Lovelock gave a lecture in California about his findings. Two young research workers, Sherwood Rowland and Mario
Molina, were in the audience and 25 years later, in 1996, they received the Nobel Prize for their now proven supposition that CFCs
could destroy stratospheric ozone.

Algae and Dimethyl Sulphide

While in Ireland, James Lovelock read a remarkable paper by the marine biologist, Professor Frederick Challenger, in which he discussed
how certain marine algae had been found to emit a volatile sulphur
compound known as dimethyl sulphide (DMS). Intrigued by that discovery, Lovelock collected various species of algae from the sea
shore and set to with his gas chromatograph. He found that nearly all
algae emitted dimethyl sulphide (DMS) but some, such as the red
hairy alga *Polysiphonium fastigiata,* were prodigious producers.

Smell of the Sea and Clouds

Not long after, Lovelock sailed on the research vessel, *Shackleton*, to
the South Atlantic with the intention of sampling the air at different
latitudes and locations. He found DMS, carbon disulphide (CS_2),
and methyl iodide (CH_3I) widely distributed over the ocean surface
and it struck him that marine life was contributing to the recycling
of chemical elements which would otherwise be depleted from the
land. The very fact that such substances were volatile meant that he
could detect their presence. In addition, it meant that they would be
carried in the atmosphere and washed out by rain as they passed over
land, so replenishing soils. Lovelock pointed out that the special
smell of the sea is none other than diluted DMS.

Although Lovelock published his results about DMS in the inter-
nationally renowned scientific journal *Nature* they went unnoticed

until the early 1980s when the US oceanographer, M.O. Andreae, confirmed that DMS was generated in sufficient quantities in the ocean to make up for any shortfall in terrestrial sulphur.

DMS Producers: Carbon Cycling and Sulphur

Apart from the red hairy alga, another prolific producer of DMS is the tiny surface-living alga, *Emiliania huxleyii,* which forms algal blooms that turn the ocean surface milky white for hundreds of square miles. *Emiliania*, like other coccolithophores, makes itself a beautifully sculpted shell — in this case, comprised of a series of hexagonal chalk plates, which give the colour. When the alga dies or is chewed up, the pieces of shell gradually sink to the bottom. Some dissolve into bicarbonate in the water, and the carbon dioxide may bubble out of the ocean. Other fragments make it to the bottom and their accumulation forms the chalk cliffs that adorn different coastlines around the world. The famous cliffs of Dover are essentially a manifestation of life's power in the cycling of carbon between the atmosphere, the oceans and the land.

While *Emiliania* is making its shells, it also releases DMS, again helping to remove the sulphur from the oceans and on to the land. Marine algae use *betaines* such as dimethylsulphoniopropionate as osmolytes to counterbalance the saltiness of their external environment. DMS is the breakdown product of this substance.

DMS, Sulphuric Acid and Clouds

Two meteorologists, Bob Charlson and S. Warren, were intrigued about the fate of DMS, and working with Lovelock they showed that as DMS evaporates from the ocean it is exposed to hydroxyl, which oxidizes DMS to sulphuric acid and methane sulphonate. Sulphuric acid droplets are effective CCNs, far more effective than sea salt. In the past, steamships emitted sulphur-laden smoke from their funnels, leaving behind a dense vapour trail, which would collect into thick clouds. Charlson, Warren and Lovelock proposed that most clouds that had formed over the ocean, particularly marine stratus clouds, came about because of DMS emissions. DMS is created by living systems, and if tropical trees are taken into acount as well, life is therefore largely responsible for rain-bearing clouds. Could life actually be regulating the climate by generating clouds which reflect away sunlight?

Since Lovelock and his colleagues suggested the life-driven formation of clouds, other scientists have confirmed that DMS is the main source of CCNs over the oceans. Meteorologists, G.P. Ayers and J.L Gras, found a strong correlation between concentrations of methane sulphonate — which is derived from DMS — and the generation of CCNs in marine air over the South Pacific. Clouds and cloudiness may help algae in the short term by bringing about localized cooling, which in turn causes local winds to ruffle the sea surface so that nutrients are circulated. Clouds may also have the long term effect of keeping global surface temperatures down and helping to generate ocean currents which are vital for the constant redistribution of nutrients. However, this is speculative. What we can say is that the recycling of sulphur, and the generation of rain-bearing clouds, are lucky side effects of the algae's struggle to keep their cells intact in a salt water environment.

Why Rain?

When water condenses in the atmosphere it produces droplets that are so small and light they remain suspended in the rising air currents. The relative humidity of air is ten times greater above an ice surface than it is above water. Based on this, Norwegian meteorologists, Bergeron and Findeison, came up with the idea that supercooled water, in the presence of ice crystals, evaporated at temperatures between 5° and 25°C, and then immediately sublimated on to the surface of the ice crystals. The crystals, as snow flakes, then enlarged and collided one with another until they had sufficient mass to begin falling. If air is warm enough as the flakes approach the ground, they melt and turn into rain; otherwise they remain as snow or hail. High flying aircraft often experience snow even though it is raining on the ground, thus lending support to the Bergeron-Findeison thesis. In the Tropics on the other hand, raindrops are swept upwards, collide and combine with other drops, until they have become big enough to fall. Both mechanisms — the ice forming and the drop forming — may, in fact, operate at the same time.

Rain Shadows

Rainfall results when two air streams meet in areas of low pressure and the warmer one is forced to rise over the other. It results too when warm, humid air from the oceans encounters a mountain barrier and the air rises and cools so that water condenses out. Once over the barrier, the air descends, gains heat and the rain ceases. For that reason, mountains tend to be wet on one side and dry on the other. Rainfall also results through convection when localized heating causes parcels of air to rise so that water condenses and latent heat is released, which maintains the upward draught until ice crystals form mixed with supercooled water. The Bergeron-Findeison mechanism then takes over. Convection provides a means whereby energy from the Earth's surface is rapidly distributed throughout the air column.

The effectiveness of the *rain shadow* in bringing desert to one side of a mountain and tropical rainforest to the other, can be seen in Colombia where the peaks of the Sierra Nevada de Santa Marta rising more than 5000 m, just 40 km from the Caribbean, make it the highest massif next to the sea anywhere in the world. Glaciers sparkle in the Sun on the summit; *paramo* — high altitude tropical alpine vegetation — then takes over as one descends. On the western side, lush cloud and tropical forest appear; on the eastern side, in the rain shadow, drought-resistant vegetation with fleshy or spiny leaves, is all that can survive. The massif then gives way to desert — the Guajira — which stretches to Venezuela.

Hurricanes, Typhoons and Tropical Storms

In one day a full-blown hurricane may unleash the energy equivalent to half a million Hiroshima-size atomic bombs. Hurricanes and typhoons form over oceans where the sea temperature exceeds 26°C for a considerable depth. They tend to form in the autumn months when the sea temperatures are at their highest, and in the trade wind belt with surface winds warming as they approach the Equator. They get their cyclonic spin from the Coriolis force and rarely form over the Equator itself. Extremely low surface pressures over a wide area and strong winds are prerequisites for hurricanes and typhoons, as well as warm water which is removed as vapour into the swirling column of air and then releases staggering quantities of latent heat when the vapour condenses as rain. The eye of the storm is an area of subsiding air some 30 to 50 km in diameter, which warms as it descends. Powerful storms run out of steam when they pass over colder water or over land, since in both instances, the amount of vapour drawn up diminishes on an accelerating scale. In general such storms last from one to two weeks.

Global warming is likely to greatly increase greatly the area of ocean with temperatures above 26°C, therefore increasing the number and intensity of tropical storms. The number of hurricanes during the autumn of 1995 suggests that warming of the oceans has already taken place.

Air and Water

Most of the water on the planet is in the oceans: less than three per cent is found on land, and more than two thirds of that in polar ice and snow; freshwater in lakes and rivers amounts to 0.1, per cent of the whole; and 0.6 per cent is partitioned between water in soil and in groundwater. A miniscule 0.001 per cent of the planet's water is in the atmosphere at any one time — therefore no more than ten days' supply and just enough if sprinkled everywhere to give each place on Earth 25 mm of rain. Water vapour is not distributed evenly: during the summer monsoon over southern Asia the air column may contain as much as 60 mm of water compared with less than 10 mm over tropical deserts. In the winter months over high latitudes, some air may be virtually dry and contain no more than one millimetre. Other times, we get literally buckets of rain: the record is 1,870 mm which fell in 24 hours on the Island of Réunion during March 1952.

The recharging of the atmosphere with water is an essential part of the *hydrological cycle*. Most evaporation and precipitation occurs over the oceans. Indeed 84 per cent of evaporation and 74 per cent of precipitation takes place over the sea. The ten per cent difference is accounted for by water vapour that passes over the land and falls there as rain. The balance is restored through run-off of water back into the sea. Most rain is not of local origin, but is brought by the wind. No more than six per cent of the rainfall over Arizona, for instance, is of local origin, and just ten per cent over the Mississippi River basin as a whole.

Dew, Frost and Fog

Air can hold a certain quantity of water vapour, depending on the temperature. Consequently, where air is in the atmosphere — whether rising or falling, expanding or being compressed, passing over the ocean or arid land — which latitude it is in, and how much is exposed to direct sunlight has profound consequences for local weather, and more generally for climate. As we know, when we walk across a meadow early in the morning following a day of bright sunshine, rain is by no means the only way in which the ground receives moisture. Dew, for instance, falls when air, that is close to the ground and therefore at constant atmospheric pressure, cools down overnight so that it becomes saturated with water vapour and the excess moisture condenses onto the ground. Ground fogs and frosts, are also consequences of clear night skies causing the land to rapidly lose its warmth of the previous day.

Smog and Traffic Pollution

Whether we see fog, dew or hoar frost, depends on ground temperatures during the night and whether the drops of water freeze, or remain in suspension. Fogs thicken when the cold layer of air close to the ground has a layer of warmer air above it. The warmer air holds down the cooler air and if the air is polluted with smoke and other industrial wastes, then the conditions become right for *smog*, and for all the health problems associated with air pollution. Cold, damp weather over London, in December 1952, led people to light coal fires. The smoke and sulphur dioxide laden fumes were trapped for days under a warmer layer of air above, resulting in thousands of additional deaths among those with respiratory problems. The 1956 Clean Air Act was brought in to prevent any re-occurrence of such devastating events.

Lessons are not necessarily learnt, more than 40 years later Delhi is located in flat plains surrounded by distant hills. In the winter, during the dry season, a typical dense inversion layer develops, with brighter, warmer air holding down the colder layer below — the same conditions that prevailed over London in the winter of 1952. Power stations send black smoke pouring into a sky from which there is no outlet because of the inversion cap. Meanwhile, unregulated traffic sends its choking exhausts into the increasingly contaminated atmosphere, with the result that Delhi has become one of the world's most polluted cities.

Clean Air

The use of smokeless fuels has markedly reduced the incidence and severity of winter smogs in Britain and other countries of western Europe. On the other hand, the increasing volume of road traffic in cities can bring on a summer smog which now contains a number of toxic substances, including high levels of ozone and peroxyacetyl nitrate (PAN) that result from photo-oxidation in bright sunlight. Los Angeles has the highest concentration of traffic in the world; it is surrounded by hills; the winds tend to be light; it has intense sunlight and, to cap it all, an inversion layer of warmer air hangs over the city for 320 days a year. The toll on human health from living in such a perpetually contaminated environment is considerable, and the damage to crops in surrounding farmland runs into millions of dollars each year.

Clean air, saturated with water vapour, can cool to temperatures well below the dew point before the water condenses. Such air is *supersaturated*. Should temperatures fall below freezing, the air may still retain its moisture without sublimating into ice. Tests on pure air indicate that the temperature can fall to as low as 40°C before water crystallizes and ice forms. Yet, the atmosphere is never that clean, even without pollution from human activities. Volcanic ash, dust from windblown soil, smoke from natural fires and industry, sulphuric acid from sulphur dioxide in the air, and even sea spray containing salts, are all present to some extent in the atmosphere. These different substances, whether particulate or simply of a chemical nature, act as CCNs and enable water vapour to condense and sublimate at temperatures close to the dew and freezing point. Such nuclei are most abundant over polluted areas such as cities, where the total may exceed one million per cubic centimetre. Smogs, such as those over Los Angeles, may form, because of pollution, when the relative humidity is actually less than the dew point, perhaps as low as 75 per cent.

Aircraft and Clouds

If road traffic causes local pollution, air traffic is more insidious because its effects are not so obviously felt and observed. Air traffic causes cloudiness, especially when the sky is criss-crossed by flight paths. In the late 1960s, in reply to a question about the environmental consequences of air traffic, the Federal German Government stated that atmospheric pollution from aircraft worldwide would at worst amount to one per cent of the total from all sources, including industry and agriculture. The matter is not one literally of quantity, but where the pollution is and what its effects are. Between 1970 and 1979 the number of air passengers doubled, and air freight increased twice as much again. A more recent study suggests that air transport produces about one third the atmospheric pollution of automobiles, which themselves account for one third of all industrial pollution.

Gisela Stief, from the Botanical Institute in Florence, Italy, has spent fifteen years studying the impact of air transport on the atmosphere. She finds that at least ten per cent of atmospheric pollutants such as carbon monoxide, unburnt hydrocarbons, nitrogen oxides, sulphur dioxide and dust, are emitted from aircraft. Worse, the emissions occur mostly in the upper troposphere and lower stratosphere where the impact of contaminants is likely to be much greater than the impact at ground level, since the gases are likely to remain at high altitudes for several years compared to barely one day when emitted from car exhausts.

The vapour trails of commercial jets drastically affect local climate and Stief finds a dramatic change in sunlight levels under the flight paths of transcontinental jets as they fly over Italy. In Vallombrosa, in Tuscany, maximum mean summer temperatures have fallen by 0.6°C compared with 40 years ago, while in Siena they have fallen by 0.8°C. The summer average hides a precipitous fall of more than 2°C during April and June, in the crucial months of fruit blossoming and ripening. The fall in temperature coincides with a ten per cent increase in cloud cover since the mid 1970s.

The number of high altitude flights is set to double again by the year 2000. A long-term study of sunlight reaching the ground over Western Germany shows a decline of eighteen per cent. In 1959, Venice received 170.56 watts per square metre of sunlight, compared to 143.93 for Hohenpeissenberg in Bavaria. By 1987, the amount of sunlight over Venice had fallen by one quarter to 131.25 watts per square metre, which was also the value found in Bavaria.

The fuel consumption of a jet plane is enormous. A jumbo jet consumes 6,500 litres of kerosene in the first five minutes after take-off, although once cruising, the rate of consumption falls to around 16,000 litres an hour. With a full complement of passengers and over long distances, the fuel consumption per kilometre is no worse than if each of those passengers was to cover the same distance overland, driving their own vehicles. The question is, do we need to fly? However good the fuel economy of a well-laden aircraft, the sheer number of flights, many million a year, has a considerable impact on the atmosphere where it is most vulnerable. The vapour trails, especially over busy routes, such as between north Europe and North America, also add CCNs which increase cloudiness. Gisela Stief has studied the skies over Tuscany in Italy since the early 1980s and has clear photographic evidence of increased cloudiness at times of the year when the skies used to be clear.

Chapter Five: Ozone Hole and Health

Skin Cancer and Ozone Holes

In New Zealand, a generation ago, it was fashionable to have a permanent suntan. No one thought twice about young children running outside naked and spending their days in the Sun. Today, if a child arrives at school without a good size hat and something to cover the neck that child will have to stay inside while his or her friends are playing outside. 'Slip, Slap, Slop!' was the message to those spending time outdoors; slip on a T-shirt, slap on a hat, slop on some sun cream. In case people do not take the warning about the dangers of excessive exposure to Sun seriously, the New Zealand media announce *burn time* along with the weather forecast. On some days, this may be only five minutes. Doctors advise annual skin checks and many people have had suspect pieces of skin excised, not least those from the previous generation for whom playing in the Sun was mandatory.

Skin cancer rates among white-skinned people living in New Zealand and Australia have soared in recent years, undoubtedly the legacy of a childhood spent uncovered in the Sun. The other reason is the sudden appearance of the ozone hole above the New Zealand sky, which is letting carcinogenic ultraviolet rays through to the Earth's surface. Not that only human beings are affected; some biologists believe that the withering and die-back now found amongst New Zealand tree species such as the cabbage tree, as well as the spreading broad-leafed *Puriri*, may be the result of increased exposure to the burning rays of UV–B that can flood through when ozone levels fall in the stratosphere.

The way our modern activities on the Earth's surface have caused the ozone hole reveals a cautionary tale, and warns us that apparently harmless industrial chemicals may be far from benign. Paradoxically, it was the relatively inert, non-toxic nature of CFCs that beguiled us into thinking that we could use them with impunity.

Ozone in the stratosphere, more than 25 km above our heads, normally absorbs UV–B and prevents most of it from penetrating to the

Earth's surface. In 1985, scientists were shocked to find that ozone was becoming severely depleted in the southern skies when the Antarctic winter was giving way to spring. When we expose ourselves to the early summer Sun in the southern hemisphere we are putting ourselves particularly at risk, since that is precisely the season when UV–B is most likely to get through.

Ozone Creation

The ozone story is replete with paradoxes. If life had not discovered how to photosynthesize and generate free oxygen, then we would not have stratospheric ozone. Without stratospheric ozone and the oxygen which engenders it, not only would we not have the stratosphere, but we would also be bombarded with extremely harmful ultraviolet radiation. Life, exposed to the Sun, may have had difficulty surviving and it is even questionable whether life would have been able to colonize the land.

Melanin, Vitamin D and Melanoma

UV–A is not only relatively harmless, but some exposure is essential for humans to build-up vitamin D in their skin. Lack of exposure to Sun can lead to young children developing rickets whereby their bones do not form properly and they may have a problem walking. Adults who keep themselves too well-covered in high latitude countries may develop *osteomalacia*, a disease which also causes the bones to soften and the legs to bow outwards. Some biologists have suggested that people in high latitudes are fair-skinned so that UV–A can penetrate to the living, outer layer of the epithelium where essential vitamin D is synthesized. On the other hand, too much vitamin D is toxic, and people with dark skins are simultaneously able to protect themselves both from ultraviolet and from excessive vitamin D. Black melanin pigment is produced by the middle layer of the skin, the mesothelium, the pigment then being carried to the dead skin of the cuticle layer. The virulent cancer, *melanoma*, is a cancer of the mesothelium.

Predicting Ozone Depletion

Both the extent of ozone depletion and the rapidity with which the phenomenon happened caught scientists unawares. When in 1974,

Sherwood Rowland and Mario Molina speculated that CFCs could cause destruction of ozone in the stratosphere, most scientists thought their contention far-fetched, even though theoretically feasible. What we now know is that many different chemical species may be involved with ozone generation and depletion.

One of the confusing factors is that ozone tends to form where industrial pollution coincides with bright sunlight. Ozone is toxic. We are therefore receiving two apparently conflicting messages: one, that ozone is increasing and causing us ill health, as well as being harmful to vegetation; and two, that it is being depleted and allowing cancer-causing UV–B through to the Earth's surface, which is also harmful to vegetation.

Nitrogen Oxides and Ozone

Paradoxically, the same chemicals are involved in both the generation and in the destruction of ozone. Nitric oxide in the presence of sunlight may cause ozone destruction. However, when hydrocarbons, such as methane, are present, nitric oxide catalyses the production of ozone. Consequently, in places of dense traffic under an intense Sun, such as Los Angeles or Mexico City, ozone is likely to build up, even to toxic levels and, at times of poor air quality, respiratory disorders increase significantly. Hence the implementation of catalytic converters which destroy nitric oxides and hydrocarbons. Ozone may be a natural component of our atmosphere, particularly in the stratosphere, but high concentrations of the gas at ground level are abnormal.

Oxygen generated on the Earth's surface through photosynthesis gradually seeps into the stratosphere, which contains most of the atmospheric ozone. Without oxygen and ozone, the stratosphere, with its clear demarcation from the troposphere would not exist. Indeed, the absorption of ultraviolet light by oxygen and ozone warms the stratosphere and accounts for the temperature rising with altitude, rather than decreasing with altitude as occurs in the troposphere. Oxygen and ozone also account for the sky being blue.

Chemical Destruction of Ozone

The destruction of ozone is helped on its way by various chemicals that, over time, are able to pass up through the troposphere and across the tropopause into the stratosphere. Before the 1970s scientists were largely ignorant of the processes by which ozone was destroyed in the stratosphere. The first hints that certain chemicals might cause problems, should they get into the stratosphere, came from laboratory work. An important lead came from the research of Harold Johnston, an atmospheric chemist at the University of California, who in the early 1970s discovered that free chlorine atoms would catalyze the breakdown of ozone. As his conclusions were based on laboratory experiments he was not sure how free chlorine, or other halogens, such as free bromine, might penetrate the stratosphere.

In 1970, even before Johnston's discovery, Paul Crutzen, a Dutch-born meteorologist at the University of Stockholm, showed that nitrogen oxides can catalyze the destruction of ozone in the stratosphere. At that time, the United States, competing with Britain and France, was considering building a fleet of commercial supersonic aircraft. Although the aircraft industry tried to discredit him, Johnston expressed his concerns that several hundred supersonic aircraft, flying at high altitudes, would release nitrogen oxides and therefore accelerate the destruction of ozone in the stratosphere. Boeing, ostensibly on the grounds of cost, gave up the idea of constructing supersonic commercial aircraft and embarked instead on developing economic long-haul jets such as the 747s. Britain and France, having forged agreements that contained heavy penalty clauses, continued with Concorde. A handful of Concordes still operates, but it is the only supersonic passenger aircraft flying.

Supersonic Flight and Ozone

Concorde produces around 40 grammes of nitrogen oxides for every kilogramme of fuel burnt. The US Space Agency, NASA, has embarked on a five year study costing $36 million to look at the impact of a fleet of 600 supersonic aircraft. Their projected impact is based on nitrogen oxide releases that are one eighth of Concorde's, and, according to NASA, would destroy no more than one per cent of stratospheric ozone. In making such an assessment, NASA has not included the effect of soot and sulphate aerosols that would also be emitted, and with which nitrogen oxides would react to form nitric acid and water vapour. The contention is that such particulate matter, combined with aerosols, could be a key factor in enhancing ozone destruction, especially when temperatures fall below -70°C, such as over the poles during the perpetual winter nights. Then, icy particles form which would supposedly capture the chlorinated compounds. With the spring Sun, chlorine would be released and able to do its worst.

Crutzen's discovery that human activities could have a significant effect on the chemistry of the high atmosphere marked the beginning of modern atmospheric chemistry. He has since helped to unravel many of the complex interactions among ozone, nitrogen oxides, hydrocarbons such as unburnt petrol from exhausts, methane and CFCs. He pioneered work on the hydroxyl radical and showed it to be of major importance as an atmosphéric cleansing agent because of its powerful propensity for oxidizing methane, nitrous oxide and sulphur compounds.

CFCs in the Atmosphere

In 1973, Rowland and Molina were working together at the University of California. They were intrigued by Johnston's discovery that ozone could be destroyed by free chlorine, and believed they had found a way in which it might be happening. Their hunch came about after hearing Lovelock talk on trace levels of the chlorofluorocarbons (CFCs) in the atmosphere. In a seminal paper, they proposed that CFCs percolated up into the upper atmosphere where, under intense sunlight, they decomposed and released chlorine. In 1974, they predicted that stratospheric ozone levels would fall by several per cent. That prediction, followed by further research, won them their 1995 Nobel Prizes.

Ozone Depletion Potential

Just as greenhouse gases have been assessed for their global warming potential, so the CFCs, the halons and other chlorinated substances, have been assessed for their *ozone depletion potential* (ODP). And just as carbon dioxide was used as the yardstick for estimating the global warming potential of other greenhouse gases, so CFC-11 is used as a yardstick of ozone (see Appendix II).

CFCs have become the best known of the ozone-destroying agents. Ironically, the unforeseen dangers of CFCs stem from the very properties which make them useful in industrialized societies. They are volatile at room temperatures, are remarkably stable, non-toxic and non-flammable. Their properties made them ideal as refrigerants and far safer to use in refrigerators and air-conditioners than toxic gases such as ammonia, or inflammable ones such as propane. They are also cheap to produce and, when they began to be used as propellants for sprays or as foaming agents for polystyrene plastics, few users were concerned with the consequences for the upper atmosphere. They were also used as solvents and cleaning agents in the electronics industry. Ultimately all uses led to CFCs escaping into the air; no one paid any attention to CFCs escaping from scrapped fridges and air-conditioners. In the United States, cars are commonly fitted with air-conditioners and their scrapping has been a significant source of escaped CFCs.

CFCs cover a range of chemicals that have chlorine and fluorine attached to carbon. In 1954 industry worldwide manufactured a total of 75,000 tonnes of CFCs; by 1974 manufacturing was up to 800,000 tonnes. Overall the concentrations of CFCs in air have increased fivefold since the mid 1970s and continue to increase by about six per cent each year. CFCs take up to forty years to decay, mainly through interactions with the hydroxyl radical and so far little more than ten per cent of that released has been oxidized.

Their oxidation leads to the production of the radical chlorine monoxide, which reacts with a single oxygen atom to form free chlorine and an oxygen molecule. One free atom of chlorine in the stratosphere will probably destroy as many as one hundred thousand ozone molecules before the chlorine is removed or inactivated. Should conditions then change, conceivably the chlorine could once again be released to embark on a new phase of ozone destruction.

It would be wrong to imagine that only man-made substances are to blame in destroying ozone in the stratosphere. Chloromethane (CH_3Cl) is primarily of biological origin, resulting from the microbial decomposition of wood as well as from forest fires. slash-and-burn farming, has been growing apace over the past 30 years and, even before the catastrophic fires of 1998, estimates attributed at least three quarters of all chloromethane emissions to the destruction of rainforests.

The Hole Revealed

In 1985, three scientists from the British Antarctic Survey, Joe Farman, Brian Gardiner and Jonathan Shanklin, discovered the ozone hole over Antarctica during the southern spring. We now have evidence that as much as fifty per cent of the total ozone is destroyed in the Antarctic region during the months of September and October,

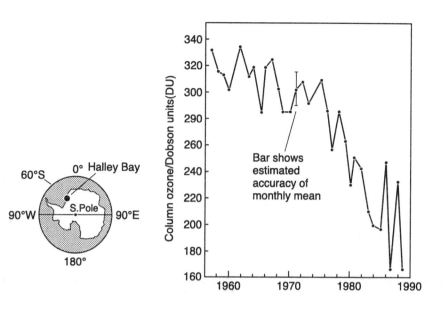

Figure 14. The October average ozone column measured from the British Antarctic Survey station at Halley Base. Before the 1970s, values were around 300 DU, but since then a rapid decline has occurred.

The units used to measure the amount of ozone in a column of air have been given the name Dobson Units in recognition of the pioneering work in the 1920s of the Oxford meteorologist, Gordon Dobson. His spectrometer enabled ozone to be measured all the way from the ground to the top of the stratosphere. Each unit has the remarkable value of 27,000 million million molecules of the gas per square centimetre and measurements of the ozone column around the globe before 1974 indicated a spread from 260 Dobson Units, close to the Equator, to 440 such units in the high latitudes. Scientists did detect some seasonal fall in the ozone column during the dark polar winters, with values falling to 280 units. Ozone then began to increase once the Sun returned with the coming of spring. In the 1960s the October (spring) values measured over Antarctica were around 300 Dobson units. However, from the mid 1970s the average October values over the region began to fall, and by 1985, had plummeted to below 200 units and have fallen even further since then.

over an area as large as the United States. The destruction is nearly complete at altitudes between 15 and 20 km. Ozone depletion is also occurring over the northern hemisphere and total losses in the upper stratosphere since 1979, amount to ten per cent. Ozone is a powerful greenhouse gas and its loss in the stratosphere has led to a fall in temperature of about 1.7°C between altitudes of 44 and 55 km.

Banning CFCs in Aerosols

The reaction to Rowland and Molina's original 1974 paper was mixed. Some scientists believed the dangers to be vastly overstated, even non-existent, while others thought that immediate action was required to forestall catastrophe. In the United States, which was then by far the largest producer and consumer of CFCs, concern focused on propellents in aerosol cans, since three-quarters of all CFC emissions were put down to this specific use. In May 1977, partly as a result of vociferous environmental lobbying, three federal agencies — the Food and Drug Administration (FDA), the Environment Protection Agency (EPA) and the Consumer Product Safety Commission —called for legislation to ban the use of CFCs in spray cans by the end of 1978.

The manufacturers maintained that the burden of proof lay with scientists to show unequivocally that the ozone layer in the stratosphere was being damaged. They felt they were being unfairly pilloried on the basis of wild speculation; how could relatively small quantities of an otherwise harmless product bring about significant changes to atmospheric chemistry more than 20 km up? Meanwhile, the critics invoked the principle that the manufacturer had to prove the harmlessness of a product before it could be licensed. In September 1976, the US National Academy of Sciences reported that the concern over the impact of CFCs was legitimate, despite the lack of hard evidence, and that non-essential uses of CFCs should be cut back.

Politics of Inaction — Models

At that time the United States' action over CFCs was considered to be exaggerated by most other countries. Government officials in the European Community, especially in the UK, advocated a policy of 'cautious inaction'. To a great extent the Europeans reacted to what was then known about chemical reactions in the atmosphere, and the models being used were simplistic to say the least. The models treated the atmosphere as a single column with no horizontal or seasonal movements of gases, and gave reassuring results in which ozone depletion caused by CFCs and other introduced chemicals was found to be minimal, especially when increases in other gases were taken into account, such as a doubling of methane or carbon dioxide. The scientists were then of the opinion that the cooling of the stratosphere, as the troposphere warmed through the build-up of greenhouse gases, would slow the chemical destruction of ozone. But, as time showed, the model was wildly wrong.

Scientists then tried somewhat more sophisticated, two-dimensional models, which took account of north-south movements of air as well as vertical ones. These forecast a global one per cent loss of ozone every decade, taking into account the increases in other gases. It also showed that ozone loss would be greatest at high latitudes during the sun-less winter months; a prediction that encouraged the modellers to believe that they had finally succeeded in achieving a measure of reality.

Yet, even these more elaborate models failed to account for the catastrophic decline in ozone discovered in 1985 by Joe Farman and his colleagues over the British Halley Base in Antarctica. It was the very cold of the perpetual night in the stratosphere that later proved to be a key factor in ozone destruction.

Political Action

In May 1981, the United Nations Environment Programme (UNEP) set up a working group to draw up a draft convention for the protection of the ozone layer. Finally, in September 1987, the Montreal Protocol was signed, committing the signatories to phasing out the manufacture and release of CFCs. Britain had tried to resist any cuts in CFC production, but facing considerable opposition at home, the government changed its mind and agreed to the Protocol. About half of CFC use in Britain at that time was as a propellant which required little hardship to find a substitute. Even, ICI, the main manufacturer in the UK, came out in favour of action.

In 1992, the Copenhagen Amendments to the Protocol came into being, which committed the signatories to eliminating the production of all halons by the beginning of 1994, and all CFCs by the beginning of 1996. Halons contain halogens such as bromine and have been extensively used as fire extinguishers. Their use in aircraft and in hospitals is to be permitted under the Protocol.

CFC Alternatives

We need something to replace the CFCs in our freezers and air-conditioners, and industry has pushed for a compromise in which the CFCs would be replaced by hydrochlorofluorocarbons (HCFCs) and hydrofluorocarbons (HFCs). The fluorine in the HFCs is apparently not a problem in terms of ozone depletion, but the chlorine in HCFCs could be a problem. However, the theory is that such substances are less stable in the atmosphere than CFCs and will mostly break down before they have time to enter the stratosphere. Given that some HCFCs would get through and cause ozone depletion, at the 1995 meeting in Vienna the industrial nations agreed to ban HCFCs by 2020, although they requested that they still be allowed to use 0.5 per cent until 2030 for servicing equipment that was still in operation.

Meanwhile, the developing countries agreed they would freeze use of these CFC substitutes in 2016 at the 2015 levels of use, and then enact a complete ban by 2040. The HFCs, on the other hand, contain no chlorine or bromine, and since they do not deplete ozone and are not included in the Montreal Protocol, their use falls outside any agreement. Both groups of substances have residence times in the atmosphere of tens rather than hundreds of years and, like the CFCs, both are potent greenhouse gases.

The global warming potentials of CFCs and related substances are indeed substantial. Even though the emissions of these compounds into the atmosphere compared to carbon dioxide are miniscule, their relative contribution to the enhanced greenhouse effect is considerable, and has been estimated at eleven per cent of the total man-made emissions and in just forty years has risen to more than one sixth of the enhanced greenhouse effect of carbon dioxide.

Today, 150 countries are party to the Vienna Convention. Ten years ago nearly one million tonnes of CFCs were produced. Now manufacture is down to less than 360,000 tonnes. Nevertheless, developing countries still have the right to produce and to use until the year 2010. This extension for developing countries was conceded at the London Meeting of June 1990, when the industrialized countries offered to set up a multilateral fund to pay for alternatives. With the fund initially set at $240 million over three years, the promise of support brought China and India into the agreement.

Political Fall-out

Obviously it is better to have something rather than nothing and the agreement by developing countries to a ban on HCFCs by 2040, and a freeze on the pesticide methyl bromide after 2002 offers some hope that ozone depletion will finally be curbed. On the other hand, up to the time of the bans countries will be allowed to expand the use of both substances. In a recent study Greenpeace maintains that if the refrigeration industry continues to grow in Asia at the rate it did over the past forty years in Japan, then by 2030, the destruction of ozone will be twice that of 1990, despite HCFCs having an ozone depleting potential one third that of CFCs.

A similar worrying situation devolves around the use the pesticide methyl bromide, which is still manufactured in Israel and the United States. Its current use is probably responsible for fifteen per cent of ozone destruction. A number of developing countries have come to rely heavily on it. Zimbabwe, for instance, uses methyl bromide to fumigate soils in preparation for tobacco and China aims to expand its use for the fumigation of vegetables and cut flowers destined for export. Irrespective of the consequences on atmospheric ozone, such countries remain determined to continue with its use.

The original Montreal Protocol only dealt with CFCs. Other important ozone depleters such as carbon tetrachloride, used in the cleaning industry, and methyl chloroform, used for cleaning and as a solvent, were not included in the Protocol, but were included in the strengthened version of 1990. Even though methyl chloroform in particular is relatively unstable and breaks down in the troposphere, enormous quantities are released, for instance, 475,000 tonnes in 1985, and enough gets through to the stratosphere to damage the ozone.

Meanwhile as the *New Scientist* reported (13 March 1999) analysis of air over the Southern Ocean has revealed that emissions of halon 1211, used in fire extinguishers has risen by 25 per cent since 1987. The emissions are fifty per cent higher than expected from manufacturing figures. The most likely source is China, which manufactures halons.

Inadequacy of Montreal Protocol

The original Montreal Protocol was grossly inadequate in that it would have led to stratospheric levels of chlorine rising as high as eight parts per billion. Scientists believe that chlorine in the strato-sphere should not exceed two parts per billion. Today, levels are at least three times that level, and show few signs of falling. Many have hailed the international agreements to halt the use of ozone-destroy-ing substances as an indication that the world can cooperate on global environmental issues. However, the international wrangle over who has been to blame, and the call for greater economic growth and development, make it unlikely that ozone depletion will be averted by the middle of the next century.

Ozone Measurements

Before 1979, ozone column measurements were made from a net-work of ground stations which included Britain's Halley Base. To get information on temperature, humidity, pressure and ozone con-centrations at different altitudes, scientists regularly released bal-loons from the ground stations. Then, in 1979, NASA launched its *Nimbus 7* satellite which carried instruments, including TOMS — the Total Ozone Mapping Spectrometer. When data from *Nimbus 7,* balloons and other satellite systems, were later scrutinized, scientists found that the area of depleted ozone extended over the entire Antarctic continent, especially between altitudes of 12 and 24 km and reached up towards the mid-latitudes.

In 1987, the United States launched a massive programme to elu-cidate the cause of ozone loss during the Antarctic spring in the months of September and October. The team of 150 scientists, oper-ating out of Punta Arenas in southern Chile, used a DC–8 and a mod-ified ER–2 spy plane to fly into the lower stratosphere and take readings of ozone concentration as well as the chemical, chlorine monoxide (ClO). At a latitude of approximately 65° S and at about 18 km up in the sky, the ER–2 discovered a sharp drop in ozone from about 2.75 ppmv to less than half that as the plane progressed to-wards the South Pole. The chlorine monoxide concentrations were a mirror image of ozone's, suggesting a strong inverse correlation be-tween the two substances.

The Significance of Carbon Monoxide

Why the interest in measuring chlorine monoxide? Its presence is di-rect evidence that ozone has been broken down. The chemical pro-cess is as follows: ozone reacts with a chlorine atom in bright sunlight, yielding the monoxide and a molecule of oxygen. UV–C, meanwhile, splits oxygen into its individual atoms, one of which reacts with chlorine monoxide, stripping away its oxygen and turning it again into a free chlorine atom. The cycle begins again, with the spring and summer Sun providing the energy to fuel the process. A single hydrogen atom will serve as well as chlorine, becoming a hydroxyl radical after interacting with ozone and then back again to a single hydrogen atom on encountering a free oxygen atom. Nitric oxide and bromine are two other potential ozone depleters and they are all present in the stratosphere, especially following our industrial activities.

Polar Differences

Meteorologists have now come up with a plausible explanation as to why the Antarctic stratosphere should have a bigger ozone hole than the Arctic. The difference between the two hemispheres gives us valuable clues as to the mechanism of ozone destruction. In effect, the story of stratospheric ozone begins over the Tropics and much of the ozone present in the stratosphere is generated above the Equator because of the intense daily sunlight and the lack of seasons. The ozone does not remain there. Stratospheric winds, driven by the spinning Earth, draw tropical, ozone-laden air towards the poles, where under normal circumstances the ozone tends to accumulate.

The destruction of ozone needs intense sunlight. During the Antarctic winter, which lasts for six months, the air becomes extremely cold and by mid-winter the Sun has vanished. We would therefore expect whatever ozone is present over Antarctica to be preserved, at least until spring. In fact, ozone levels do not build up during the winter, for the simple, but crucial reason that rapid cooling over the polar stratosphere during the late autumn and winter draws in powerful westerly winds — the circumpolar vortex— which swirl around the pole to about 60° S. This vortex of rushing air remains in force until some time in November when the Antarctic is in the midst of its summer. It is that polar vortex which prevents stratospheric air moving all the way south from the Tropics.

During the Antarctic winter nothing much happens, at least chemically. But then, once the Sun starts rising again over the continent, the chemicals that can break down ozone wake up from their chilly, perpetual night state and begin to do their worst.

Cold Temperatures, Water and Methane

Even the more elaborate models failed to account for the extent of ozone destruction, and scientists realized that they had left one essential part of the process out of their calculations. The extreme cold of the circumpolar vortex, with temperatures falling below -80°C, is part of the story; the other part has to do with water. A relatively new phenomenon has begun to manifest itself in the stratosphere — the visible appearance of translucent icy clouds that can be seen in ghostly form at night. Originally, before the 1980s, the Antarctic stratosphere was thought to have next to no water, so where did those *noctilucent* clouds spring from? Paul Crutzen believes the answer lies in the increased emissions of methane from the Earth's surface.

Methane concentrations in the atmosphere have been increasing by about one per cent per year since the 1950s, which we can blame on increased cattle ranching, especially in the American Tropics, rice paddy production in South-East Asia, tropical forest destruction through slash-and-burn farming, and leaks of natural gas from pipelines, not least from Siberia. Some of that methane percolates into the stratosphere, where, again in the presence of intense sunlight, it is oxidized to water and carbon dioxide. Crutzen maintains that methane is therefore the prime source of those beautiful noctilucent clouds.

The supposition is that the bitter cold of the sunless winter not only forms the clouds but draws chemicals such as CFCs to them, which have percolated into the stratosphere. The ozone-destroying chemicals concentrate there, biding their time until the Sun returns and melts the clouds. With the melting of the clouds, the mixture of ozone destroyers is free to do its worst and the destructive cycle is set in motion.

Why should the ozone be more vulnerable over Antarctica than over the Arctic? The air above Antarctica gets considerably colder during its winters than the Arctic air, and the circumpolar vortex over the Arctic Circle is less well-defined so that ozone from the Tropics can get through more easily and make up for any losses. Nevertheless, the situation is deteriorating over the Arctic, and more ultraviolet penetrates to the ground today than it did 20 years ago. One concern is that the loss of ozone will lead to the Antarctic stratosphere getting colder, because ozone is a greenhouse gas. Consequently the circumpolar winds could become stronger and effectively isolate the Antarctic stratosphere from the Tropics for longer, making the depletion even greater. In December 1987, when the Antarctic summer was at its height the stratospheric temperature above Britain's Halley Base was 15°C colder than normal. Another consequence is that the colder stratosphere will lead to a warmer troposphere because more ultraviolet radiation will get through. The result of that would be to compound global warming.

If nothing else, the flurry of scientific activity after the discovery of the hole has illuminated a fascinating world in which life's processes are seen to be intricately involved with the chemistry of the atmosphere. Here we see how disparate activities — growing rice and raising cattle or using CFCs in air-conditioners and freezers — can come together to cause major damage to a natural system. None of this was predicted thirty years ago when Lovelock first measured a few molecules of CFCs in the Irish atmosphere. The lesson is there: we cannot tamper with natural systems and expect nothing to change. Nor can we expect these changes to be beneficent.

Sun Exposure

Our preoccupation with the ozone hole comes about for the selfish reason that we fear for our health. Exposure to the Sun is potentially more dangerous now than it was half a century ago. You have only to stand outside in the Sun in New Zealand to feel its burning, penetrating quality. Science indicates that a one per cent loss of stratospheric ozone leads to a two per cent increase in the amount of UV–B which reaches the Earth's surface.

Cancer and the Ozone Hole

Dark-skinned people derive their colour from the black pigment, melanin, which to a great extent protects the skin from UV–B. Even though ozone is generated over the Tropics where the Sun is more intense, it tends to disperse and concentrate over the poles. As a result as one moves closer to the Equator more UV–B gets through putting fair-skinned people at risk. Epidemiological studies in the United States of death among white males from the virulent skin cancer, melanoma, show a significant correlation as one passes from the high latitudes towards the Deep South. In Quebec, for example, with a latitude of 46° N, the annual average death rate from the disease between 1950 and 1967 was around five per million people. In Texas, the death rate was around twenty-two, and in Florida nineteen. On the basis of such studies, the US Environment Protection Agency believes that every one per cent decrease in the ozone column could lead to a one per cent increase in deaths from melanoma, and a three per cent increase in other skin cancers.

Currently melanomas kill an estimated 6,000 people a year in the United States alone. Other skin cancers, although some thirty times more common, are more likely to respond to treatment. Consequently, the death rate is similar to that from melanoma. In Britain, the National Radiological Protection Board (NRPB) has also voiced its concern about the dangers of a thinning ozone layer over the mid-latitudes of the northern hemisphere. Over the past fifteen years cases of malignant melanoma have doubled in the UK, and it now accounts for one out of twelve cancers among people in their 20s and 30s. Meanwhile, since 1988, the NRPB has found a five per cent increase in ultraviolet radiation getting through to its headquarters in Oxfordshire, which could translate into a considerable increase in melanomas and other cancers for those who expose themselves to the English Sun. To make us worry a little bit more, the NRBP, thinks that UV–A may not be as harmless as was previously thought. Sunlamps use UV–A, and some sunscreens are also unreliable. The danger too is that those who use sunscreens are likely to spend more time in the Sun. Meanwhile, an increase in exposure to UV–B is likely to lead to an increase in eye problems such as cataracts.

Effect on Marine Life

How will other living systems cope with increased ground level ultraviolet radiation? Life is vulnerable, especially when its delicate cells have no thick covering nor special pigment to absorb the radiation before it can penetrate the inner cytoplasm. Plankton in the sea or delicate algae in fresh, clear water are particularly at risk. If the water is clear, ultraviolet can penetrate to at least 20 m; if murky, then it may penetrate no more than 5m. Biologists believe that all off-shore life, including fish and coral reefs may be at risk from increased exposure to UV–B. Juvenile forms of plankton, including the larvae of fish, crustaceans and molluscs, are especially vulnerable. The Southern Seas, around Antarctica, are one of the richest regions for marine life. In the past, little UV–B penetrated because of the high concentrations of ozone in the stratosphere. However, Antarctica, as well as the Arctic, must now be at grave risk from the sudden and dramatic decline in ozone and the inevitable increase in UV–B radiation during their respective springs. We will certainly feel the impact in terms of declining fish yields.

The Land and Ultraviolet

Terrestrial plants have evolved with ultraviolet radiation and are probably better protected than marine plankton. Nevertheless, of 300 cultivated food species that were tested, two thirds have shown sensitivity to UV–B in terms of impaired growth. Photosynthesis in particular has declined, causing stunted growth and smaller yields of fruits, berries and seeds. Wide-ranging ecological effects are likely to follow as more vulnerable species give way to UV–B resistant ones. Tests on Loblolly pine trees exposed to UV–B radiation, resulting from a twenty then forty per cent depletion of ozone, show dramatic stunting as more UV–B gets through. Recent research on maize has brought to light another hazard of increased exposure to UV–B, that of 'jumping *mutator* genes' in which genes that are currently stable start 'jumping' from one site on the chromosone to another, where they cause disruption of the plant's development and fertility. Even though ozone depletion may have originated from the activities of a few industrialized nations, the problem has clearly become global in its implications.

Chapter Six: The Oceans

WATER is essential for life; as a medium for dissolving salts and gases such as carbon dioxide and oxygen; as a liquid that transports substances to and from the cytoplasm of cells; as a constituent of organic molecules. These remarkable properties enable water to play a critical role in the processes of climate and make this planet habitable. In addition to dissolving minerals, transporting them to the oceans, lubricating the sliding of one tectonic plate over another, water carries energy around the globe, distributing it as part of a planetary conveyor belt system.

The energy carried by the world's water parallels energy distribution in the atmosphere. The ocean absorbs the lion's share of solar radiation and transfers much of that energy to the atmosphere through the evaporation of water. Energy is also transported through oceanic currents. The Gulf Stream, in a flow of water equivalent to more than one hundred Amazon rivers, carries more than one thousand terawatts of energy from the Equator to high latitudes. That equals the energy required to evaporate the rain that falls over the Amazon, and is one hundred times the entire energy consumption of human societies across the world. The carriage of energy is in the form of heat, and the waters moving northwards are on average 8°C warmer than the cold, deep waters travelling south.

Oceanic Transport of Heat Energy

Climatologists estimate that the oceanic transport of energy in the subtropics and middle latitudes may amount to one quarter of the total energy transferred across the planet. As we have seen the atmosphere is heavily implicated in energy transfer, particularly over the Tropics where cumulonimbus clouds push moisture-laden air to great heights at considerable speeds. Between 70° N and 70° S nearly all energy transfer from the oceans to the atmosphere is by means of latent heat; the heat required to turn water into vapour. In the higher latitudes and polewards, the transfer of energy is through the movement of warmer air from lower latitudes. We now know that on average some 1800 equatorial thunderstorms develop each day, sufficient to carry energy in the form of latent heat that accounts for the transfer from the Equator to the higher latitudes.

Ocean Currents and Freshwater Run-off

The total amount of water in all its various forms of snow, ice, vapour and liquid on the planet amounts to 1,384 million cubic kilometres (1 cubic kilometre = 1 billion tonnes). 97.4 per cent is sea water, and eighty per cent of the remaining freshwater is in the form of snow and ice. Perennial ice covers eleven per cent of the land surface and seven per cent of the oceans in the form of sea ice, which may increase to ten per cent during the polar winter months. Every year 3,300 km³ of freshwater run off the land into the Arctic Ocean, adding thirty centimetres of freshwater to the surface and substantially reducing its salinity. This freshwater is a critical component of the process that drives the major ocean currents. It does so by affecting the freezing point of the ocean, first raising it because of the lower salinity, and, leaving saltier waters behind when it freezes which tend to sink. In effect, the freshwater flow impinges on the rate and timing of the sinking of the surface waters. We are now coming to understand that subtle changes in the amount of freshwater running into the oceans around Antarctica and the Arctic can cause fundamental switches in ocean currents and therefore abrupt changes in climate.

The Antarctic continent occupies an area of fourteen million square kilometres, nearly all of which has permanent ice, in some places 2,400 m thick. The area with ice is seven times larger than that covering Greenland in the northern hemisphere. Ice and snow have high albedos, on average reflecting as much as seventy per cent of sunlight back into space. Ice therefore contrasts with sea water, which absorbs most of the sunlight that strikes it. The summer melting of sea ice and its transformation into water therefore dramatically alters the amount of sunlight absorbed or reflected. Subtle changes in the rate and timing of the oceans' thawing and freezing therefore govern how much sunlight is absorbed at the Earth's surface or reflected away.

Water and Albedo

Water's greatest impact on the planet's albedo is when it is condensed into thick clouds. Clouds account for 82 per cent of the total albedo of the planet. With the addition of the water's surface area in all its forms, and its respective albedos, the Earth owes nearly 94 per cent of its albedo to water. The relative warmth of the sea surface as well as the continents makes a considerable difference to the amount of cloud, and therefore to the amount of sunlight penetrating at sea level.

Computer simulations of the state of the Earth at different phases of its existence indicate the importance of albedo. A small shift in albedo, for instance, and the models suggest that the planet could become irretrievably frozen. Yet, we have evidence that throughout its 4.5 billion year history the Earth never plunged into a frozen state from which it could not emerge. M.H. Hart, who carried out simulations of that kind in 1978, assumed that as the Earth cooled it lost its original atmosphere and the basis of our present atmosphere came about through gases, such as carbon dioxide, methane, ammonia and water vapour, being pumped out via volcanic activity, and then degassed from the Earth's mantle. He also accounted for the Sun gradually becoming more luminous over time. As a main sequence star, the Sun gradually heats up, on account of the accumulation of helium, which as a greenhouse gas, retains heat, thereby speeding up its thermonuclear transformation from hydrogen. Today the Sun is emitting about 25 per cent more heat and light than it did 4.5 billion years ago, when the Earth first began forming.

Hart began his simulations with certain assumptions about the early composition of the Earth's surface and then let the model run through successive stages until arriving at a simulation of the Earth in which the atmosphere, albedo and greenhouse effect, mirrored the reality of today. The model would therefore only possess credibility if it arrived at the correct finishing point.

With an atmosphere dominated by carbon dioxide, water vapour, methane and ammonia, the model indicates that the greenhouse effect was initially large. The average surface temperature 3.5 billion years ago might have been as high as 42°C, compared with 15°C today. However, the high surface temperature would have resulted in a cloud-covered Earth, which reflected a considerable proportion of the Sun's energy. Then, starting some two billion years ago, oxygen began to accumulate in the atmosphere and the concentrations of gases such as methane and ammonia fell precipitously. The greenhouse effect was much reduced and the surface temperature fell to 8°C. For the first time polar ice caps could form. With lower surface temperatures and

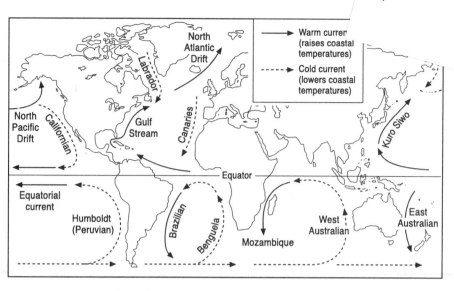

Figure 15. Major ocean currents.

less vapour in the atmosphere, clouds were sparser and more sunlight could penetrate, but now it encountered ice and snow at the poles and was reflected away. In his simulations, Hart found that should the surface temperature fall to 0°C then, despite volcanic activity, the Earth would be irreversibly stuck in a frozen state.

Clearly, the planet has somehow achieved a delicate balance between overheating and overchilling. In the simulation series that gives today's atmospheric composition and the correct surface temperature, as well as radiation from the Earth to space, the maximum variation in surface temperature over the past 2 billion years has been calculated at 10°C. The actual temperature range was found to lie between 5° and 15°C, with a mean surface temperature of 10°C.

Ocean Currents

As the atmosphere has giant convection cells which distribute energy and moisture, so the seas of the world have their own circular currents, known as *gyres*. For the most part the upper waters of the oceans are set in motion through friction between fast moving air and the more sluggish surface layers of the ocean. However, when ıe wind driven surface waters encounter a continental land mass, they collect and are then deflected, making a circular loop, clockwise in the northern hemisphere and anticlockwise in the southern hemisphere.

The trade winds, through their persistence and regularity, drive the most powerful of these currents — the subtropical gyres —from east to west. However, the Atlantic and Pacific Oceans are bounded on both sides by continents and, on reaching land to the west, the oceanic waters turn and make their way back across the ocean, this time pushed on their way by the westerlies. The Gulf Stream in the North Atlantic, and the *Kuroshio* situated in a comparable latitude in the North Pacific are the result of these surface currents. Both carry warm water to a depth of up to one hundred metres, at rates of up to 8 kilometres per hour, from the south-west to the north-east across their respective oceans. The Gulf Stream alone carries more than one hundred times as much water as all freshwater rivers and streams on the Earth together. And, because the North Atlantic, offers a clear run to the Arctic Circle from west to east, the Gulf Stream actually carries its warm waters far to the north, in what is known as the *North Atlantic Drift*.

The Gulf Stream and the nature of its gyre is a major factor in the temperate climate of northern Europe. Cornwall and western Scotland pride themselves on their palm trees which grow in such a northerly clime, courtesy of the Gulf Stream. Without that constant, unchecked flow, temperatures over all northern Europe would plummet and no palm tree would have a chance unless grown in a greenhouse. The Kuroshio current, on the other hand, is hemmed in by the meeting of the Asian and American continents at the Bering Straits, and its movement is curtailed by the North Pacific current which sweeps down from the Arctic. Nevertheless, it does for Seattle, Vancouver and Alaska what the Gulf Stream does for Europe.

The westerlies also drive one segment of the subtropical gyre which circles towards the polar region and is returned westwards through the polar easterlies. Narrow gyres also operate either side of the Equator, the trade winds pushing the equatorial waters westwards with the returning flow moving eastwards in the *doldrums*. The land masses impinging on the Arctic prevent the polar easterly from circumnavigating the Arctic Circle, whereas the clear separation of the Antarctic from other continents allows the Antarctic current to be truly circumpolar, and is known as the *West Wind Drift*. The integrity of this current prevents warmer waters penetrating and as a consequence high latitudes in the southern hemisphere tend to be far colder than comparable latitudes in the northern hemisphere. Here we have a parallel with the circumpolar vortex which is significantly stronger than its northern counterpart.

The direction of the flow in a gyre, clockwise in the northern hemisphere and anticlockwise in the southern hemisphere, is a consequence of the Earth's rotation and the Coriolis force, which in principle pulls waters to the right of the winds in the northern hemisphere and to the left in the southern. As well as generating the gyres, the winds also dissipate their energies, shaping the water into waves.

Waves may be no greater than ripples or they may be more than 5 m high, measured from the trough to the crest. The largest wave ever recorded had a height of 34 m. One wave striking the shore can generate a shock wave of up to thirty tonnes per square metre. Choppy waters in confined bays have short wavelengths, whereas waves formed out in the open ocean have longer wavelengths. Despite the surface waters being highly disturbed, the waters at depths of half the wavelength remain almost still.

Tides

The combined gravitational pull on the Earth of the Sun and Moon generates the tides. The Moon stabilizes the Earth's rotation and without it the Earth would tumble chaotically in its trajectory around the Sun. That would have put an end to established seasons and the Earth could have finished up as an iced-up planet, incapable of sustaining life. The tides, with their ebb and flow on the shoreline, first dousing a community of organisms in water and then leaving them to dry out until the next round, may have played an essential part in giving life the necessary environment in which to adapt to an existence on land. Life had to learn how to resist desiccation, and this it did early on by forming semi-permeable membranes and sheaths. The blue-green cyanobacteria were able to protect themselves from daily tidal exposure and still resist bombardment from ultraviolet that, without an ozone layer and a clearly demarcated stratosphere, could reach the Earth's surface unchecked.

The Moon, on account of its proximity, exerts more than twice the tidal force of the Sun, so that whichever side of the Earth faces the Moon has the highest tides at that moment. On the other side of the Earth, the tides will also be high. Consequently, since the Earth's rotation means that a particular point of its surface faces the Moon once every twenty-four hours and fifty minutes, each place has two high tides and two low tides in just over the length of a day. When the Sun and Moon are aligned the tides facing them are at their highest. These are *spring tides,* and occur every fourteen days. When the Sun and Moon are at right angles to the Earth, the tides are at their lowest. These *neap tides* also occur every fourteen days.

Different places on Earth, depending on ocean depth, coastline shape and the angle of the Moon and Sun relative to the Equator, have different ranges between high and low tides. The Mediterranean has a miniscule tidal range of one centimetre, the English Channel one of two metres, and the Bristol Channel and the Rance Estuary in Brittany a tidal range as great as 12 metres. The Bay of Fundy, with a tidal range of 18 metres, has the largest tidal range in the world.

Upwellings and Nutrients

Not only do surface currents transport energy, but they carry minerals and nutrients that support marine life. The top one hundred metres are heated by the Sun, to 10°–15°C in middle latitudes, and over 20°C in the Tropics. Below that depth, not only are the waters relatively still, but the temperature begins to fall rapidly. This fall — the *thermocline* — continues to depths of around one thousand metres, but from then on down, irrespective of latitude, the temperature remains nearly constant at around 5°C. The thermocline acts as a barrier to the mixing of lower waters with surface waters. In hot tropical waters, the thermocline is extremely effective in keeping the waters separate; in the middle latitudes, especially during the stormy winter months, the thermocline loses its integrity as a distinct boundary layer and mixing takes place between the surface and deeper waters. That mixing replenishes the surface waters with nutrients for the spring and summer plankton blooms.

Deep ocean currents, far more sluggish and massive than the surface currents, are driven by differences in density between one body of water and another. In general too, the deep waters move in the opposite direction to surface currents. However, and crucially for marine life these waters do mix. Sinking occurs primarily in the polar regions, where wind blowing off the ice cap freezes the surface waters. Ice forms from freshwater and consequently the chilled, unfrozen water becomes saltier and therefore more dense. This water, initially colder than the deep ocean waters, begins to sink and on reaching the ocean floor flows from the polar regions towards the Tropics, where it wells up and completes the cycle. The deep waters carry oxygen and nutrients and consequently refurbish the surface waters.

By measuring the ratio between different carbon isotopes in sea water, scientists are able to determine with some accuracy just how long deep waters have been out of contact with the atmosphere. On average deep waters take between 200 and 1000 years before welling up to the surface. Evaporation removes water from the surface of the oceans, and calculations of the evaporation rate across the globe suggest that the average time water spends in the oceans is about four thousand years. Overall, water recirculates between four and twenty times before it evaporates.

A longer recycling of the ocean takes place through *subduction*, at the boundary between an oceanic and a continental plate. Approximately one cubic kilometre of sea water a year is drawn down into the Earth's crust. A seemingly negligible amount, but, over geological time, such a drawdown gradually removes minerals such as sodium and chlorine from the sea. One billion years would probably have to pass before half the mineral content of the oceans was removed in this way. But minerals are always being replenished through run-off from the land, at rates faster than their removal by tectonic processes. Evaporite basins, in which portions of the sea are sealed off from the main body of water and gradually lose their water through evaporation, are undoubtedly a far more important mechanism for taking otherwise highly soluble salts, such as sodium chloride, out of general circulation. Although the composition of the seas has changed greatly over geological time, the seas have never become so salty to make it virtually impossible for life to survive. Could life have taken a hand in eliminating salts from the sea through its painstaking construction of barriers, such as the Great Barrier Reef off the Australian coast? Certainly, most if not all of the marine salt deposits now found on shore appear to be enclosed in limestone basins, limestone that was laid down by living organisms millions of years ago.

The upwellings of cool deep water occur where surface water is driven offshore by winds. The cold, nutrient-rich waters of the Antarctic, for instance, surface along the western coast of South America, replacing the surface waters that have moved westwards because of along-shore southerly winds that find passage between the Pacific Ocean and the high ridge of the Andes. The trade winds assist in this shift of surface waters by driving them towards South-East Asia.

Strange Properties of Water

Usually when substances solidify, they become denser and sink, yet ice is lighter than water, and by surfacing is more likely to melt than if it accumulated at the bottom of the ocean where it would not be exposed to the spring and summer Sun. The Earth would be a very different place if ice were denser than water and sank, as the oceans would then gradually freeze from the bottom up and only a thin surface layer would remain liquid; the hydrological cycle would literally grind to a halt as water would no longer be transported in updraughts to higher latitudes. Life would barely exist, if at all. Fortunately for us, water has unique characteristics that set it apart from any other compound. In fact, by floating and presenting a glass-like surface, which reflects sunlight, ice provides a counterpoint to the light-absorbing properties of the sea. A dynamic between cooling and warming can therefore develop which offers the possibility of climate regulation.

The Gulf Stream

The sinking of water in the North Atlantic and its flow downwards to the Equator and beyond towards Antarctica is crucial to the northwards flow of the Gulf Stream. Should the cold waters fail to sink, then the Gulf Stream would be blocked much further south than is currently the case, and the warm waters that now bathe the western reaches of Britain and even penetrate into the North Sea, would be replaced by cold waters flowing south from the Arctic Circle. Winter temperatures would plummet, possibly by as much as 10°C in little over a decade.

Some disturbing findings, admittedly from climate models, warn us that we are heading for a catastrophic seizing up of the Gulf Stream on account of global warming and that such an event may not be in the distant future. Vittorio Canuto, a physicist at NASA's Goddard Space Centre in New York, together with his colleagues, find that as little as 0.25 per cent more freshwater flowing into the North Atlantic from melting glaciers in Greenland and Northern Canada will bring the northwards flow of the Gulf Stream to a shuddering halt. Should equivalent carbon dioxide levels in the atmosphere rise to four times their pre-industrial levels — a likely event if the trends of growing industrial emissions continues — then the Gulf Stream, again according to the model, will be permanently shut down, this time because of insufficient cooling of the surface waters.

To make such concerns all the more real, the Arctic is showing all the signs of a planetary warm-up. Polar bears, for instance, are feeling the brunt of the warming as the once-continuous ice is now breaking up, leaving them stranded. Greenland, too, is reflecting its Viking name once again as the ice shrinks from the coastline and grass ruggedly pushes its way up from the sparse soil.

Like Canuto, climatologists at the Met Office's Hadley Centre have modelled the flow of the Gulf Stream under different global warming scenarios to determine how much, if at all, the conveyor belt circulation of warm tropical waters to the high northern latitudes would stall if carbon dioxide levels rose at the rapid rate of two per cent a year, then stabilized at four times the present concentration. The model shows that the strength of the Gulf Stream circulation will decline sharply by one quarter. With a growth in carbon dioxide levels as assumed by the IPCC's business-as-usual scenario, the decline in the circulation sets in around the turn of the current century and in a matter of thirty years falls to one third of its current level. That decline represents a substantial loss in energy transfer. We are therefore talking about a loss to the British Isles and Northern Europe of thirty times the energy used by all humanity. Although contentious, the Met Office climatologists claim that such a loss will be more than offset by the warmer temperatures that accompany the direct effects of global warming:, and according to them, temperatures over northern Europe will still rise.

A salutary warning as to how abruptly a switch can occur comes from the recent discovery that from one year to the next the currents in the Mediterranean have undergone a complete about-turn. In the past, cooler waters from the Adriatic flowed along the seabed in an easterly direction towards the Aegean Sea and the Levantine coast. The Adriatic waters were replaced by the westerly flow of warm water from the Aegean. Now, the warm waters of the Aegean, instead of their westwards flow, are now sinking rather than remaining on the surface, and are flowing eastwards. The system has flipped.

Wolfgang Roether of the University of Bremen puts the blame for the flip on increased evaporation, because of a warmer climate, and a sharp decline in the amount of freshwater flowing in, because of increased urban use and the construction of dams, on rivers such as the Dnieper, Nile and Danube. This has led to the surface waters becoming much saltier and therefore more dense than the underlying waters. Those warm, salty waters are now sinking, stalling and even completely reversing the circulation that presumably prevailed until now. The current switch indicates the potential impact of global warming and the uncertainties we face in the future from abrupt climate change.

Although climatologists at other institutes in the United States and Continental Europe all agree with the general principle that global warming will cause a critical change in the flow of the Gulf Stream, differences have emerged in the degree to which stalling occurs under a global warming regime. Paradoxically, the saltier waters flowing back into the Atlantic from the Mediterranean could keep the conveyor belt going. According to Eelco Rohling of the Southampton Oceanography Centre, the waters from the Mediterranean flow deep below the surface, northwards up to the Faroe Isles where they rise, mix and then sink rapidly, drawing down surrounding water, including those of the Gulf Stream. The more salty the waters from the Mediterranean the stronger the pull. This process would seem to counter the potential seizing up of the Gulf Stream in its northern stretches. Yet the more rapid melting of Greenland's ice sheets could intervene and sweep away the saltier water. Again, we would be in line for a slowing down and even seizure of the Gulf Stream.

Recent research from the Goddard Space Centre indicates that Greenland's ice sheet is melting faster than it is being created and is losing as much as a foot a year from its surface. That means an increase in the freshwater flow into the Arctic Ocean and suggests we are heading for the very scenario outlined by Canuto and the Hadley Centre.

Fearful Prospect

Climate models, matched to evidence derived from ocean sediment cores, indicate three different modes of North Atlantic circulation: one, a *warm* conveyer belt mode as has operated over the past 10,000 years. Two, a *glacial* conveyer belt mode which operated during the past ice age; it was shallower and did not extend further north than the south of Iceland. Three, a *weak* conveyer belt resulting from large amounts of melt water capping any circulation by forming a surface lens of freshwater. This last mode is one that climatologists fear could be repeated through global warming generating more dilute and warmer surface waters.

Stefan Rahmstorf of the University of Kiel in Germany, has identified another mode, also the result of a large influx of freshwater into the North Atlantic, in which the conveyer belt remains vigorous, but with the sinking taking place much further south than is currently the case. The evidence is that whenever the Gulf Stream stalled, or was pushed south, northern Europe was pitched into a cold spell. According to Rahmstorf, by disrupting the conveyor belt, we could be triggering calamitous cooling throughout Europe. As he states: 'The consequences for ecosystems, agriculture and society could be severe.'

Past Episodes of Cooling

Sudden reversals of climate, from cold to warm and back again seem to have been the order of the day until about 12,000 years ago, when the Earth gradually moved into its current interglacial regime. Even so, the climatic record from the end of the ice age until the present has, as we have seen, alternated between cool and warm periods. Reid Bryson and his colleague B.M Goodman find geological evidence that periods of enhanced volcanism may have played a significant role in precipitating planetary cooling that lasted for periods of several hundred years or more. One period of cooling took place about 12,000 years ago. Geologists describe this last fling of the most recent ice age as the *Younger Dryas*, on account of an attractive yellow-petalled tundra flower that greatly extended its range at that time as a result of the cool, dry conditions that spread from the edge of the ice shield. The Younger Dryas was a period of volcanism, as were other periods, such as those known as Cochrane, Indus and Vandal, and the effect of each of these episodes, according to Bryson and Goodman, was to set in train an extended period of cooling, as seen in the build up of ice in Antarctica and the advance of glaciers in New Zealand. Methane concentrations from ice cores taken from Greenland also show a significant drop, during such periods, by as much as a third, indicating a drying out at the Tropics and therefore correlating with the sudden fall in temperature.

Summer Chills

A small change in energy input appears to have considerable consequences. As Reid Bryson remarks, in northern Canada, in the district of Keewatin, snowbanks may not finally melt away until the end of July. Less than a month later the first snows may arrive. Were the summer to shorten by just three weeks then new snow would pile on old and a new round of continental glaciation could be in the making. As usual paradoxes abound when it comes to diagnosing what conditions would most likely instigate a period of glaciation or of melting. Mild winters tend to be associated with stormier weather and heavier precipitation and, as a consequence, mild winters and cool summers are more likely to lead to accumulations of snow than are cold winters and hot summers. Computer models of past climates confirm that periods of continental ice melting, as occurred 15,000–6,000 years ago, take place when the winters are colder, to be followed by hot, dry summers.

Bryson believes that periods of sudden cooling are more likely to be a result of increased volcanic activity rather than changes in the solar constant seen by the rise and fall in the number and frequency of sunspots. Why not the two together? Possibly volcanism and changes in the solar constant did at times coincide to accentuate climate change. Nevertheless, once some change had been set in motion, it could bring about a cascade of events, with repercussions way beyond the initial trigger. It is in this context that we should take Canuto's model seriously, especially as we are now gathering the historical evidence to back it up.

Initially, the discovery of momentous and rapid swings in climate posed something of a puzzle, since scientists had previously believed that climate changes were slow, gradual affairs. Once we knew about ice ages, from the work of Louis Agassiz in the past century, we assumed that the planet was either in an ice age or coming out of one; the last ice age was considered to be no exception. Some recent discoveries have rudely awakened us to the idea that climate is far more transient, far more dynamic, than we had ever imagined. This very transience is suggestive of a system that is far from chaotic: on the contrary, it is a system in which the energy received by the planet is being constantly re-distributed in order to bring about a momentary equilibrium.

Oceans as Indicators of Past Climate Change

In 1988, Hartmut Heinrich, of the German Hydrographic Institute in Hamburg, began looking at deep sea sediment brought up from the bottom of the north-east Atlantic. That sediment corresponded with the last ice age and by looking at fossil material, he was able to correlate dates with temperature and salinity. For instance the calcite shells of zooplankton known as *Foraminifera* are tell-tale indicators of species, and different species prefer particular temperatures and salinity. The relative proportions of oxygen–16 to oxygen–18 in the fossilized shells also indicate temperature and salinity. Colder water retains the oxygen–18 isotope more than warmer water and consequently a higher proportion of the heavier isotope indicates that lower temperatures prevailed at the time that the shells were formed in the surface waters. Such data indicates that the temperature of the north Atlantic has risen by 10°C since the last ice age.

Glacial Slip

Among the samples he gathered in from the ocean bottom, Heinrich discovered coarse pieces of rock that were far too large to have been blown there by winds, and he surmised that they must have been carried on the backs of icebergs. Fossil samples showed that surface waters at the time when the rock fragments were deposited were between 1° and 2°C colder than during the preceding period and that salinity was down by as much as ten per cent. Heinrich came to the conclusion that vast chunks of the North American ice sheet must have slipped into the North Atlantic and been carried to where cold polar waters encountered the warm waters of the Gulf Stream. There, they began to melt.

Rocks carry their own magnetic signature and the magnetic susceptibility of the sediment on the ocean floor gives a clear sign of how much rocky material the icebergs were carrying and what their routes were. Today, the Gulf Stream and the North Atlantic Drift keep Arctic icebergs pinned to the far northwest Atlantic. The sediment samples studied by Heinrich indicate that the encounter between the polar waters and the Gulf Stream was further to the south.

Heinrich concluded that sudden meltings of large chunks of the glacial ice cap were associated with changes in the North Atlantic currents and the Gulf Stream in particular. Wallace Broecker at Columbia University and others, such as Nick McCave, Nick Shackleton and Harry Elderfield at the University of Cambridge, independently confirmed Heinrich's findings, but they came to the surprising conclusion that the ice had come from the interior of North America rather than from sheets of ice that were breaking off at the edges. Nor was such an event a one off: the sediment layers, with their fossils, interdispersed with debris carried by the ice, indicated that similar events occurred several times over.

Sediment sampling indicates that *Heinrich events* could take place within a matter of years, with dramatic implications for climate, at least in the northern hemisphere. In each such event the equivalent of as much as half the ice now present in Greenland's ice sheet would have slipped into the Atlantic. Heinrich believed that the events as he described were caused by the continental ice expanding over Greenland and North America, with the result that massive pieces broke off at the edges and floated off into the ocean to be carried away by the currents until they melted. We can get some idea of what it might have been like from the present-day collapse of pieces of the Larsen ice shelf in Antarctica.

Ice Sheets and Rocks

Heinrich's suggestion of a *calving* (shedding) of the ice at its edges might make logical sense, but it does not fit the facts. When John Andrews, a glaciologist at the University of Colorado, studied the sedimentary material from the ocean floor he found that some of the rock fragments were of a type that could have come only from the centre of the North American continent. Andrews suggests that the ice cap overlay loose, broken, unstable sediment. As the ice sheet built up it began to exert ever greater pressure on the ground and trapped geothermal heat. The two phenomena combined, the pressure and the heat, melted the base of the ice and the entire sheet began to slide, gathering momentum and surging through the Hudson Strait into the North Atlantic.

Douglas MacAyeal, of the University of Chicago, has modelled such events and finds them not only plausible, but has even been able to time one event after another. In essence, the natural build-up of ice, its subsequent melting and slide, followed by another period of build-up, takes approximately 7000 years. The evidence from deep sea cores confirms the timing of such sequences.

The Heinrich events that have been studied to date all took place within the last ice age. The presumption is that the circulation pattern of the North Atlantic would have swung between the two extremes; one where, as today, circulation is intact and the Gulf Stream can advance right up to the Arctic Circle; and the other in which it is blocked far to the south by a stalling of cold, sinking, saline waters. The North Atlantic would therefore fluctuate between warmer and cooler phases: warmer when the flow of freshwater left behind sufficient salt water to cool and sink, and cooler when the flow of freshwater, derived from the melting of continental ice, left the ocean surface with too little salt water to sink.

Finally, once the ice sheet began shrinking and the flow of freshwater went down, Heinrich events would no longer occur and, over time, the mass circulation patterns of the North Atlantic would be established as we know them today. But, could global warming trigger Heinrich events from the substantial ice sheet that still sits over Greenland, or similar events in Antarctica? The evidence is that the Greenland sheet did not take part in the Heinrich events, probably because it was too well grounded on a solid rock base. On the other hand, if global warming triggered a sea level rise such that tongues of ice from the continental ice shelf were pushed upwards, that process could loosen the ice sheet and ease it on its way into the ocean. The melting of the sheet would raise sea levels still further and the process would accelerate into a full-bodied Heinrich event. Should that happen, it would undoubtedly affect oceanic circulation patterns and the sinking of surface waters. The transfer of heat from the Tropics to the high latitudes would decline, setting off dramatic changes to global climate.

Ocean Current Flip

Other surprises are probably in store for us in the Antarctic. The pattern is similar to that in the North Atlantic. Surface waters will not sink until ice forms, when the sea water left behind becomes saltier and correspondingly more dense. Meanwhile, warmer water lies underneath the surface, but it cannot break through because of a thin layer of water called the *pycnocline* which acts as a barrier. Only when the ice has come during the winter are the surface waters dense enough to break through the pycnocline. That rupturing of the pycnocline is like the opening of the flood gates and the warmer waters can now burst through to the surface. The deep water has travelled

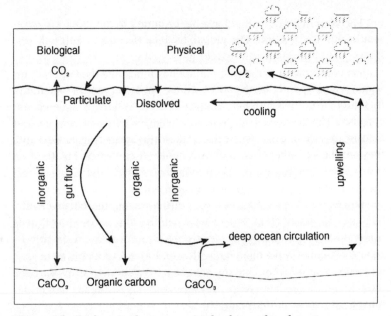

Figure 16. Carbon in the oceans in the form of carbonate.

down from the Tropics and even the North Atlantic, and being warmer than the surface water, on surfacing it releases both heat and carbon dioxide. Then, having exchanged its heat and gases with the atmosphere, the waters sink again to become part of the global oceanic circulation.

Quite what will happen to that circulation and to the release of heat and carbon dioxide with global warming is not known, but it could make a substantial difference to energy and gas uptake into the world's oceans. A straightforward expectation would be that with global warming the surface ice would melt and warmer water would still break through, but scientists are not so sure that would be the outcome. They point out that with global warming atmospheric humidity will increase which translates into more water vapour going aloft. That would result in increased snow over the continent, and the snow melt in the summer months would make the ocean less salty. The surface waters would then lack both the density and the low temperature to break through the pycnocline. The deeper waters would fail to break through and the seas would stagnate. The Weddell Sea in Antarctica is a sea that never loses its covering of sea ice. A strong possibility is that the ocean surrounding Antarctica would become like the Weddell Sea, never losing its ice. The spread of ice and its retention throughout the year would definitely affect the global balance of energy through reflecting sunlight back into space.

The oceans are a key player in climate change, providing both insulation and stability from perturbations, but at the same time, likely to switch from one mode to another in terms of large scale ocean currents. The oceans are also the site where Earth's crust is generated from volcanically hot regions, the mid-ocean ridges. Equally, they are the site where ocean plates encounter continental plates and in forcing them up, are themselves pushed down. The rising up of the continents, with their mountains and volcanoes, has major consequences for climate. Exposed rock surface is an immediate candidate for the weathering processes whereby rock is dissolved and mineralized. That process depends on carbon dioxide in the atmosphere which, in washing down with rain, forms a mild, but nonetheless effective, acid that eats at the rock. Weathering takes carbon dioxide out of the atmosphere and therefore has an impact on the atmosphere's content of greenhouse gases.

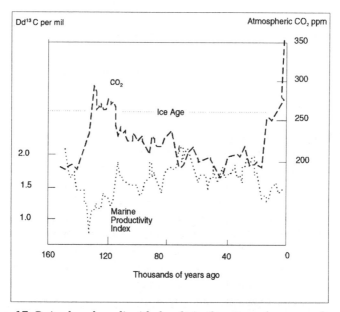

Figure 17. Raised carbon dioxide levels in the atmosphere correlate with a precipitous decline in plankton activity 140,000 years ago. This stage is followed by higher primary biological productivity with an associated drawdown in carbon dioxide.

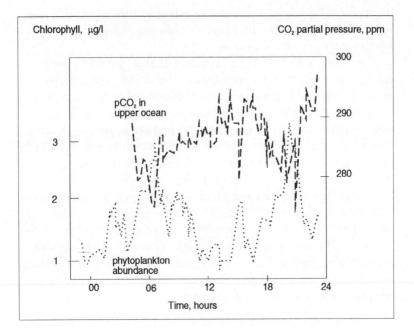

Figure 18. Carbon dioxide drawdown into the ocean is clearly associated with enhanced phytoplankton activity.

Sinks and Sources

The oceans, with fifty times more carbon (dissolved as carbonates) than in the atmosphere, are both a *source* and *sink* for the greenhouse gas, carbon dioxide. As the oceans warm, the solubility of carbon dioxide falls and the gas is expelled. Global warming therefore accelerates emissions of the gas, increasing global warming even more. In contrast, a cooler ocean will take up more carbon dioxide, thereby diminishing its greenhouse effect.

However, plankton influence the rate and amount of carbon dioxide that is drawn down from the air. They can speed the process up by at least ten times and, as Lovelock points out, operate better when the oceans are cooler rather than warmer, primarily because more nutrients are then accessible to them.

The evidence for phytoplankton activity is startling. Satellite shots reveal an impressive fifty to one hundredfold range of chlorophyll concentrations in the oceans, indicating where high levels of photosynthesis are taking place. Not surprisingly the regions with the highest concentrations of chlorophyll overlie precisely the same areas of ocean where carbon dioxide uptake is at its highest, as in the North Atlantic. Oceanographers believe that if it had not been for the activity of marine biota the atmospheric levels of carbon dioxide in the immediate pre-industrial era would have been 450 ppmv rather than the actual level of 280 ppmv.

Disruption of oceanic currents and in particular the Gulf Stream conveyor belt would bring about formidable self-enhancing feedbacks. If the population of phytoplankton begins to crash because of a combination of warmer surface waters, a curtailing of the conveyor belt, and by exposure to ultraviolet from the hole in the ozone layer, then inevitably much less carbon dioxide will be drawn down into the ocean depths: hence more warming. The build-up of carbon dioxide in the atmosphere because of our industrial and agricultural emissions means that the surface waters are enriched in CO_2 by more than 40 micromoles/kg compared with those of a century ago and at least four fifths of the carbon dioxide derived from fossil fuel burning and the destruction of forests is still to be found in the upper 750 m of the great subtropical ocean gyres.

Sea Level Rise

For the past 1500 years the sea has risen at the rate of about ten centimetres a century: that rate has now virtually doubled. With a doubling of CO_2 from pre-industrial times to approximately 600 ppmv, the UK Met Office predicts a rise of 44 cm by 2080, from the expansion of water alone. The Met Office points out that if carbon dioxide levels increase by one per cent a year for the next seventy years and then stop, sea level rise from thermal expansion will continue to increase by as much as seventy centimetres over the following five hundred years, long after we should have stopped emitting greenhouse gases.

This rise does not include any melting of the ice over Antarctica or Greenland. Should the West Antarctic Ice Sheet melt in its entirety that would add as much as 6 m to current sea levels: the Greenland ice sheet would add another 6 m, an event that could be set in motion over the next century should warming proceed as rapidly as forecast in the more aggressive business-as-usual scenarios of the IPCC. Whereas several years ago scientists claimed that it would take a phenomenal amount of surface warming over Antarctica to destabilize the ice sheet, opinions are now changing.

In an international ministerial meeting on Antarctica organized by the New Zealand government in January 1999, Peter Barrett, a geologist at Victoria University in Wellington, warned that the entire West Atlantic sheet was becoming unstable and could soon break away.

Effects of Rising Seas

Small, low-lying islands, such as atolls will be the first victims of global warming as the expanding seas rise and swamp them. Already some islands in the Maldives have become uninhabitable. The cost to save such islands will be disproportionately high given the relationship between land area and coastline. The problem is not solely sea level rise. Other factors, such as spring tides, heavy rain, deep depressions at sea and storm force winds can turn what appears to be a minimal sea level rise into a catastrophe.

The Chinese coast is particularly subject to high sea levels because of strong depressions and storm force winds. During typhoons, surges of up to 5 m are not unusual. Some parts of the world are sinking, for instance around the Black Sea and parts of Indonesia. Such subsidence is generally at a rate of around 30 cm a century, although much higher rates of several metres have also been known. Indonesia has fifteen per cent of all the world's coastlines and as much as forty per cent of its land-surface is vulnerable to rising sea levels.

Salt Water Intrusion

Salt water intrusion is already a serious problem in the mouth of the River Rhine, penetrating as much as fifty kilometres upstream. In the Netherlands, according to Gerrit Hekstra in the Ministry of the Environment, it might become necessary to flood reclaimed land with freshwater from the Rhine, just to keep sea water at bay. In the UK too, the government has decided to abandon the age-long struggle to protect vulnerable coastlines, such as along the Norfolk Coast. Instead, the idea is to allow salt marshes to re-establish themselves as the primary barrier against sea rise and surges.

Meanwhile, several of the UK's nuclear power plants, such as Hinkley Point and Sizewell, are vulnerable to sea level rise and storm-force sea surges. Sizewell A, for instance, is sited a few metres above sea level in an area where sea levels are rising at twice the national level. Worldwide, many other nuclear power stations are sited on the coast, some in areas that may be threatened by sea level rise.

Nuclear power stations, particularly in the United States, now keep all their nuclear waste in cooling ponds on site, within the containment structure of the reactor. Those ponds need active cooling twenty four hours a day every day. Any threat to the reactor from sea level rise is therefore magnified because of the presence of such waste. Chemical waste dumps in low-lying areas, such as Pitsea in Essex, could also be swamped because of a rising sea combined with sea surges. Such flooding would spell catastrophe for groundwater supplies as a result of toxic chemical contamination.

Vulnerability to Rising Seas

The world's coastline is between 0.5 and 1 million kilometres long. A one metre rise in sea level would affect up to 5 million square kilometres, therefore three per cent of the total land area of the planet, but more significantly as much as thirty per cent of the total cropland in the world. The loss of land that would go with a sea level rise of several metres would be simply catastrophic.

The Met Office estimates that with coastal protection remaining as it is but the sea rising, in the mid 21st Century as many as 78 million people worldwide, especially in South East Asia and the small island states, could be at risk from wild storm conditions. The claim is that with *evolving protection* the numbers of people at risk would be reduced by 28 million to 50 million overall: but all that is without taking any Antarctic ice melting into account. The number at risk would rise tenfold or more were the sea to rise at double the rate indicated from today's climate models.

According to Mark Meier of the University of Colorado, we have already seen as much as a 4 cm rise over the past fifty years because of melting glaciers. Even a one metre rise would threaten many of the world's major cities, such as New York, London, Bangkok. Six metres would be devastating. The West Antarctic Ice Sheet is grounded below sea level and is already showing signs of instability. Massive chunks of ice, covering some thousands of square kilometres, are now breaking off and melting, as they drift northwards into warmer waters. In the mid 1990s the Larsen A ice shelf toppled and broke away. It was some 8000 square kilometres in surface area. However, that was nothing compared to Larsen B, which in early

1998 showed signs of following suit and, with a surface area equivalent to twice the size of Norfolk, would be the single largest iceberg to be spawned over the past fifty years. Antarctica appears to be warming faster than anywhere else on the planet and grass is now beginning to push up from what was frozen wasteland just a few years ago. Penguins, in particular, are suffering from the heat and a number of their breeding colonies are now threatened with extinction.

Meanwhile, in the Alaskan Arctic, Inuit communities are becoming increasingly worried at the signs of warming all around them. The ice is melting, the tundra is drying, summer rainfall is significantly less and the winters markedly warmer. Melting permafrost is threatening communities with unprecedented landslides and storms at sea are getting wilder. All such signs have been predicted in the IPCC models, but not for now — for fifty years time.

The consequences of sea level rise are unthinkable, particularly if the ice sheets begin melting and slip into the sea. Major capital cities, such as Bangkok may have to be abandoned, or protected at great expense through a system of massive dykes and barriers. The Netherlands is a past master at coastal protection, but with every metre rise in sea level the cost of protection increases disproportionately. There is already great cause for concern, not only because of the threat to cities, but also because a large proportion of the world's best agricultural land will fall victim to the waves.

Chapter Seven: Rocks, Soils and the Origins of Life

NEARLY thirty years after Christopher Columbus' voyage to the Americas in 1492, the Portuguese explorer, Ferdinand Magellan, set out to circumnavigate the globe. Having rounded the straits that now bear his name, he pressed on to the Philippines. There, in a melée with natives, he was killed, but some crew escaped and, in the one remaining ship of five, sailed around the Horn of Africa and so back to Spain. For the first time we had living proof that the Earth was round and that the continents could be placed, albeit somewhat crudely, one after the other around a spherical globe.

Despite the fundamental change to our concepts of the planet that came with circumnavigation, we still believed then that the planet was as it always had been, unchanged since the time of creation. However, as exploration continued, some strange discrepancies came to light: on the one hand, plants and animals seemed to vary substantially from one part of the planet to another, elephants in Africa or Asia, being quite distinct from polar bears in the Arctic. On the other hand, explorers and natural historians were discovering some extraordinary parallels and similarities. On his voyage around the world in *HMS Beagle* the young Darwin collected rock and fossil specimens from the Falkland Islands in the South Atlantic which were found to bear a close resemblance to specimens that had been gathered in Cape Province of South Africa by Andrew Geddes Bain. It was hard to escape the conclusion that a great land mass had once existed south of Africa, of which the Falklands was part.

Half a century later, in 1910, Robert Scott set out on his ill-fated expedition to Antarctica in which he and his men perished. Despite the struggle to survive, Scott refused to jettison the sample of rocks and fossils, and one found with the men's bodies, was of the plant, *Glassopteris*, which later provided essential evidence that the continents had once been grouped together as a super continent.

Shifting Continents

The idea of continental drift and later of plate tectonics, coupled with volcanism, has revolutionized our thinking about the origins and evolution of life, from micro-organisms such as bacteria to multicellular organisms like ourselves. Our views about climate have certainly been transformed because of geological sciences and we now see indissoluble links between the atmosphere, the oceans and the shifting continental plates.

Earth as Superorganism

The Greek and Roman philosophers certainly saw processes at work on our planet which wrought change as well as continuity. In his work *De Rerum Natura (On the Nature of Matter),* Lucretius remarked that '... the sum of things is always being renewed.' Two thousand years later, in the late eightennth century, James Hutton, a Scottish doctor and scientist, believed that complex physical and chemical processes had been involved in the formation of the Earth as we know it, which could be unravelled through the observation of volcanic eruptions, erosion processes and the aftermath of earthquakes. He viewed the present as the key to the past, now known as *uniformitarianism*, where the origins of rocks and landscapes could be traced to the processes which actually created them.

William Harvey's discovery, nearly a century before, of the blood circulation in the body inspired Hutton to see parallels with the systems that governed the Earth. In 1785, the discoverer of carbon dioxide, Dr Black, gave a lecture to an Edinburgh scientific society on behalf of the ailing Hutton, in which he spoke of the Earth as a *superorganism*, and that scientists should undertake a physiological approach to understand how it regulated itself.

Like the circulation of the blood, the waters of the planet carried dissolved gases and minerals from one part of the globe to another, while the winds acted like the lungs, inhaling and exhaling air. The mountains, meanwhile, like bones, provided both solid structure and the source and place of storage of those same essential minerals. And, just as an organism regulates its metabolism and keeps its structure and form, while taking in nutrients and excreting waste, it appeared that the Earth had also devised ingenious ways to maintain a flow of minerals across its surface.

Hutton's ideas, though crude, nevertheless anticipated Lovelock's Gaia thesis of the 1970s. Lovelock later invoked an alternative name, geophysiology, for his thesis, in which living organisms and the Earth were seen to be inseparable, part of a self-regulating system that managed to restore and maintain an equitable climate in the face of internal and external shocks.

Cataclysm and Catastrophe

Every so often, planetisimal fragments are large enough to escape annihilation as they hurtle at tens of thousands of kilometres an hour through the atmosphere. When they strike they do so with the power of thousands of the largest hydrogen bombs. The impact may be big enough to set the Earth ringing like a bell and to send such a mass of debris into the atmosphere that the Sun is cut out for months at a time. The scarred face of the Moon, shows signs of many impacts, as does the Earth when you look for them; the Canadian shield of north Saskatchewan and Northwest Territory stands out on account of the myriad of lakes scattered over the surface, as does the Manicougan Crater in Quebec.

On account of plate tectonics and the constant weathering of the surface, the Earth tends to cover up the ravages of the past. Yet, craters aside, other tell-tale signs include deposits of minerals, such as nickel alloys that have been left behind in sufficient concentrations to make mining worthwhile at Sudbury, Ontario.

The precious metal, iridium, is a rare mineral on our planet, but has conceivably been brought here by a planetisimal. The discovery of a thin layer of this metal, at various sites across the planet, is indicative of a massive explosion. Luis and Walter Alvarez were first to find the iridium sandwiched between a deep layer of limestone and a layer of sand and mud, marking the boundary between the Cretaceous period one hundred million years ago, when the dinosaurs were very much alive, and the Tertiary period (the K/T Boundary), 65 million years ago, by which time those prehistoric creatures had mysteriously vanished.

Demise of the Dinosaurs

The Alvarezes speculated that their discovery of iridium, coincided with the demise of the dinosaurs, pointing firmly to the notion of the Earth being struck by a planetisimal the size of Mount Everest that would have released the energy equivalent to millions of atomic bombs. The evidence of a cataclysm exists in the abrupt disappearance of minute coccolith shells that, over time, had generated the limestone of the Cretaceous Ocean, and the sudden appearance of a layer made up of fossil-free muds. Carbon dioxide levels would have risen several fold and, once the debris began to wash out of the sky, have taken surface temperatures up by an average 10°C. The K/T boundary was also a time of greatly increased volcanic activity, conceivably the shattering blow on the Earth's surface having spawned a myriad earthquakes and volcanoes, the relics of which are to be found in the Deccan plains of India.

The effects of that cataclysm lasted up to one million years and life in the form of complex, multicelled organisms must have barely hung on. As James Lovelock emphasizes, the apparent recovery of the Earth from such a destructive event is evidence indeed that life is the active agent in cleaning up the atmosphere and, re-establishing a balance between the output of carbon dioxide and its drawdown for photosynthesis, in relation to greenhouse gases and surface temperature.

In 1992, ten years after they suggested that a planetisimal had been the prime cause of the K/T boundary, the Alvarez team found evidence of a massive crater at Chicxulub, on the edge of the Yucatan peninsula in Mexico. The geology and timing was right, the event having occurred 65 million years ago. Surrounding the crater was striking evidence that the impact had been swiftly followed by mountainous tsunami — the destructive waves that generally ensue after major earthquakes and volcanic eruptions occur in the bedrock under the ocean.

The Restless Earth

The environment is constantly changing; nothing is ever quite the same, least of all climate. And it is those concepts of change, combined with uniformitarianism, that provided the basis for theories of continental drift and plate tectonics. Together with the discovery of radioactivity a century ago we now have an underlying energy source for the emerging geological ideas of the restless Earth.

In 1912, Alfred Wegener, a German meteorologist, came up with his theory of continental drift in which he proposed that the planet's land masses were once joined together in a supercontinent — now known as *Pangaea* — leaving the rest as ocean. 200 million years ago, Wegener's Pangaea began splitting into two large continents, Laurasia, which went north, and Gondwanaland in the south. Later these two large masses began splitting up further, some fragments of one colliding with fragments of the other in a process that still goes on. The Indian subcontinent, for instance, came adrift from between what is now South Africa and Antarctica, and then shifted northwards to collide with the southerly moving part of Eurasia. This collision continues today and the Himalayas are still rising, heaved up as the two continents force up against each other.

The world's other imposing mountain range, the 7000 km long Andes, was formed as the American plate thrust up against the Nazca plate of the Pacific. Lake Titicaca, at an altitude of 4000 metres, was once part of the ocean, but as the mountains rose on either side, it was carried up, to become the highest salt lake in the world. The rising up of the Andes also reversed the drainage flow out of the Amazon Basin, accounting for the shallow fall of less than one hundred metres along several thousand kilometres of the Amazon River, across the Amazon Basin to the Atlantic Ocean. Equally it gave rise to the sunken area in the central part of South America. This region, known as the Pantanals, is famous for its swamps.

Evidence of Continental Drift

Wegener's evidence for continental drift rested on the discovery in Brazil and South Africa of a fossil reptile, *mesosaurus*, that lived in the Permian period, 250 million years ago. Another intriguing discovery was of a plant that lived in the Carboniferous period, 345 million years ago, and which was found only in Antarctica and India. Wegener also noted that certain rock formations bore close similarities to others in different continents, such as the Appalachian Mountains of the eastern USA and the Caledonian mountains of Scotland.

Initially Wegener met with considerable opposition to his theory, mostly because no known mechanism could account for the fragmentation of Pangaea. Then, in 1948 a discovery was made that clinched the matter of continental drift: the Mid-Atlantic ridge, a continuous mountain range along the ocean bed, was found to be made up of young, volcanically generated rocks in which the history of their emergence was encapsulated in their magnetism.

The Atlantic Ridge

The extraordinary finding was that magnetic polarity appeared to switch suddenly, and when the basalt on each side of the ridge was dated the switches coincided closely in terms of age and distance from the central ridge line. The only reasonable explanation was that the magnetic field not only fluctuated, but every so often completely reversed. Analysis of the magnetic polarity of the Atlantic oceanic crust either side of the ridge indicates that the so-called magnetic north has reversed completely more than 170 times over the past 76 million years. Studies completed in 1962 showed that the Atlantic was spreading and widening by up to 5cm a year, so that in a million years North America will be fifty kilometres further away from Europe. In 200 million years, a continental plate would be carried half way around the world.

New Crust and Continents

Iceland was born from the mid-Atlantic ridge. In January 1973, the crust literally tore apart for several kilometres at Helgafell, letting out lava and ash, which enveloped and destroyed hard-won farmland. Nearly ten years earlier, on 14 November, 1963, the crew of an Icelandic fishing boat felt an underwater explosion and then saw smoke, steam and ash burst out of the sea. A volcanic cone emerged 130 m from the seabed and within six months was covered in hard, crustal lava — now the island of Surtsey. Iceland is home to at least one third of all volcanic lava emitted on to the Earth's surface over the past 500 years. Mid-ocean ridges, which mark the boundary between diverging plates, are where three-quarters of all lava ejected from the mantle finishes up.

Geology and Climate

Geology and climate are closely interwoven. Through volcanism, carbon dioxide is pumped into the atmosphere, where it stays before being flushed out and caught in a chain of events that may finish with it being drawn back, via plate tectonics, into the volcanic process, and so out again, maybe tens of millions of years later. The volcanic part of the carbon cycle is by far the slowest, on account of it being the least likely fate for the carbon. Most carbon is recycled many times over throughout the atmosphere, the continents and oceans. During cooler phases in the Earth's climate, such as an ice age, as much as 200 billion tonnes of carbon may be drawn down from the atmosphere. The same amount in the form of carbon dioxide will equally return to the atmosphere during a warm period.

The oceans, with fifty times more carbon (dissolved as carbonates) than is in the atmosphere, are both a source and sink for the greenhouse gas. As the oceans warm, the solubility of carbon dioxide falls and the gas is expelled. Global warming therefore accelerates emissions of the gas, increasing global warming even more. Vice versa, a cooler ocean will absorb more carbon dioxide, thereby diminishing the greenhouse effect of the atmospheric gas.

At the same time, plankton influence the rate and amount of carbon dioxide that is drawn down from the air. They can speed the process up considerably and, as Lovelock points out, operate better when the oceans are cooler rather than warmer. One reason is that colder oceans exhibit more upwellings, bringing up more nutrients.

Pangaea and Warming

The form of the continents, packed together as Pangaea, or dispersed into today's configuration, makes a profound difference to climate, affecting the circulation patterns of air masses and ocean currents alike. The configuration may have altered sea levels and the amount of atmospheric carbon dioxide, which itself will have changed the energy balance at the Earth's surface.

By far the most important natural system for removing carbon dioxide from the atmosphere is where the gas reacts with the silicate in the Earth's surface crust. The more the crust is exposed, through the lowering of sea level and the thrusting up of mountains, the faster this *weathering* process can proceed. However, should carbon dioxide be removed from the atmosphere faster than volcanoes can replenish it, then the rate of weathering will decline, until a balance is reached between the two processes. In principle, the carbon dioxide washes out of the sky as the mild acid, carbonic acid, which eats away at rock and releases silica and various carbonates and bicarbonates, as well as forming clays such as kaolinite. All these minerals are soluble to some extent in water and gradually wash away into the oceans.

A Continental Cycle

The continents have gone through a series of stages of coming together and then breaking up, with immense consequences for the dissipation of heat from the Earth's surface. The driving force for this supposed aggregation and dispersal is the heat welling up from the mantle and the rate at which it can be removed. Overall, the oceans dissipate six times more heat than the continents. The accumulation of heat beneath the supercontinent causes it to rise and because this exposes more surface, the weathering of rock and its erosion can take place faster, thus gradually taking carbon dioxide out of the atmosphere and helping to reduce the greenhouse effect. In the end a supercontinent builds up enough geothermal heat beneath the surface to break the land mass apart and send the various plates on their way.

When the continents are most spread out, then the heat tends to leave them and they subside relative to sea level, thereby diminishing rock weathering. In the next stage, the continents tend to clump together again so allowing the completion of the cycle. Heat then builds up once again. Models indicate that the entire process takes approximately half a billion years.

Life comes into the process through accelerated weathering and the drawing down of carbon dioxide, thus cooling the planet surface and, especially if ice caps and glaciers develop, causing a drop in sea level, which itself exposes still more land. By means of photosynthesis and respiration, life therefore acts as an important intermediary in the exchange of carbon between the atmosphere, soils and ocean.

If the amount of carbon dioxide returning to the atmosphere at any moment was perfectly matched by that being withdrawn, then the cycle would be complete and relatively unvarying. Yet, we have only to look back over the evidence of the past 600 million years to see that the levels of atmospheric carbon dioxide have at times been nearly twenty times today's levels, and were four times higher 200 million years ago. London University botanist, Jenny McElwain, has found that the numbers of leaf stomata go up in plants exposed to low levels of carbon dioxide and down when the carbon dioxide levels increase. She has correlated her findings with geochemical data and suspects that fifty million years ago, carbon dioxide levels were actually double those today, perhaps as a result of enhanced volcanic activity.

Carbon Cycle

Life on land holds approximately as much carbon as in the atmosphere, and soil double the amount. The other main source of carbon is fossil fuels: coal, lignite, petroleum and natural gas. The estimated recoverable resources of fossil fuels for industrial use amount to approximately 2000 billion tonnes of carbon — enough to last 400 years at current rates of consumption. Imagine the consequences of our drive for industrial development should we put several thousand billion tonnes of carbon into the atmosphere over the next couple of hundred years. We wouldn't just be facing a doubling of atmospheric carbon dioxide content, but a tenfold increase, especially as other sources of carbon are driven out of the ground through widespread global warming, such as methane in permafrost.

Trapped Methane

Considerable carbon is trapped under permafrost in the form of *clathrate*, a complex of methane and water. Clathrate is found mainly in tundra regions and buried in northern continental shelves. Gordon MacDonald, a geophysicist from the MITRE Corporation in the United States, believes the warming up of the permafrost regions, when an ice age gives way to warmer weather, initiates the release of methane which accelerates warming so that even more permafrost melts (see Chapters 3 and 6) The added pressure of a higher sea level on the continental shelves could also squeeze out methane. The exact proportion of methane ascribed to clathrate, found in Vostok and Greenland ice bubbles, and how much to methane from decaying bog vegetation is unknown. Should the current phase of global warming continue and push back the permafrost, then conceivably, clathrate could start gushing out of the ground and, with the release of methane, push greenhouse gas concentrations up still further. The Earth could then find itself caught in a runaway phase of global warming.

Peat Bogs and Climate

Not all feedbacks lead to warmer conditions; some feedbacks go the other way, and induce global cooling. One in particular, involves bogs and peatlands, and their tendency, especially towards the Arctic Circle, to take over from all other types of vegetation. Lee Klinger, a biologist at the US National Center for Atmospheric Research in Boulder, Colorado, believes that *Sphagnum moss* bogs may even have been critical factors in the development of ice ages.

There are four main factors: wet, marshy conditions; potentially acid soils; surface freezing and, because of fogs and mist, the maintenance of a high, sun-reflecting, albedo. Bogs tend to form on low-lying land once covered by oceans, such as large areas of the Canadian north as well as of Russia, amounting to 5 million km². The soils contain relatively large concentrations of sulphur of marine origin, either in the form of sulphides or sulphates. When faced with lack of oxygen in the presence of sulphate, certain species of sulphur bacteria reduce sulphate to sulphide. At the surface a different group of sulphur bacteria operates, which converts sulphide back into sulphate. The soil's acidity now rises substantially.

Bog plants, such as *Sphagnum*, can tolerate acid soils; they also absorb up to 40 times their dry body weight in water, so keeping local conditions wet; and they tend to form a hard pan layer beneath them that prevents drainage. As *Sphagnum* grows it lays down organic matter, which is prevented from decaying by the acidity and anoxic conditions of the soil. The net result is that peat gradually accumulates. Klinger estimates that worldwide between 500-860 billion tonnes of organic carbon in the world are now held as peat — a significant amount when pitted against the 750 billion tonnes of carbon in the atmosphere.

Peatlands develop their own misty, damp, atmosphere which encourages dew and fog in a self-perpetuating wet cycle. That all-year-round dampness prevents drying out and drought conditions, while the mistiness, even in summer, doubles the albedo compared to a nearby forest. In the winter, snow covering raises the bog's albedo to eighty per cent compared with about fifteen per cent over a conifer forest. The higher albedo contributes considerably to the maintenance of cool conditions. Once the process is set in motion, the peat bog becomes self-feeding and expanding.

Cooler conditions certainly favour the advance of peatlands, and as the mosses take over they cause further cooling through an increase in surface albedo and a drop in greenhouse gas concentrations in the atmosphere. Snow and ice begin to accumulate and the temperature drops further. The expansion of glaciers and the ice cap inevitably destroys peatlands, but that destruction is probably balanced by the advance of peatlands into the mid-latitudes. Also, as sea levels fall more land is exposed and so more bogs develop. At their peak sedge moss bogs may have buried as much as one thousand gigatonnes of organic carbon with a subsequent significant fall in average surface temperatures.

Fires and Global Warming

With the help of sedge moss it is conceivable that the planet could enter a perpetual ice age — a feedback from which there would seem little escape. How then can the planet break out of this freeze in climate? The chances are that the very success of sedge moss bogs is their undoing. In the equation of carbon dioxide drawdown and the burial of organic carbon, oxygen is released into the atmosphere and its concentration therefore rises. Higher oxygen levels mean that fires are easier to start and more difficult to extinguish. Forest fires replace the carbon that has been drawn down out of the atmosphere, so that carbon dioxide levels once again begin to rise. As James Lovelock has pointed out, fires may be one mechanism by which life is able to achieve some sort of balance between a perpetual ice age and an increasingly hot surface temperature, as well as a mechanism by which atmospheric oxygen remains close to 21 per cent.

Another factor would be the movement of ice and glaciers. As they advanced from the Arctic Circle, glaciers would also bear down on the peatlands, and once the rate of destruction exceeded the spread of peatlands, the balance between drawdown and emission of carbon dioxide would be subtly changed. With carbon dioxide levels rising in the atmosphere and increased surface temperatures, sea levels would begin to rise, both the result of warmer temperatures as well as from freshwater flow from melting glaciers. The peatlands of the continental shelves would become swamped, causing rapid decomposition of organic matter and the emission of methane.

We have evidence from gas samples in ice cores that methane levels rose in a spurt at the transition between the glacial and interglacial periods. That sudden increase in methane concentrations in the atmosphere, coinciding with the breaking out of an ice age, suggests that the scenario described for the swamping of land through sea level rise, may be relatively close to the truth.

Meanwhile, much of what we are doing today in terms of industrial activity is running counter to the glacial/interglacial pattern that appears to have established itself over the past few million years. Not only are we burning fossil fuels, but we are flattening forests and draining wetlands thus adding considerably to the production of both methane and carbon dioxide. We are also pushing back the permafrost boundaries, potentially releasing vast quantities of methane from clathrate. Without question, we are in the throes of precipitating warming, while simultaneously putting out of action those natural checks and balances that have served in the past to maintain some sort of equilibrium.

Chapter Eight:
Gaia — Climate and Life

Gaia Theory

Our perspective of life has undergone a rude awakening. In his *Origin of Species* Charles Darwin put forward the idea of life being engaged in a ruthless struggle to adapt to a hostile environment. But another idea has now begun to take hold, of the planet Earth being a dynamic system in which life plays a fundamental role. James Lovelock's *Gaia Theory* is undoubtedly the most far-reaching scientific theory we have of life's extraordinary influence over the surface of our planet; including the atmosphere, oceans and continents, volcanism and plate tectonics, since they are all part of the dynamic interchange that goes on in what the nineteenth century engineer and geologist, Edward Suess, called the *biosphere*. The Earth's climate is therefore not simply the result of inexorable physical processes, such as the stream of energy we receive from the Sun.

By altering the chemistry at its surface boundary life actually transforms the nature of the atmosphere, oceans and even the surface of rocks. What is a microscopic or molecular change wrought by one organism, be it bacterium, fungus or multicellular organism, becomes a momentous transformation when multiplied a billion times over.

The annual cycle of carbon dioxide between vegetation and the atmosphere, as seen in the polygraph from the Mauna Loa laboratory in Hawaii, is a conspicuous example of the consequences of such bio-magnification. If the concentrations of greenhouse gases are unique to this planet, at least in our particular solar system, then we have life to thank for that, and for altering the albedo of our planet so that heat and light are differentially absorbed. Climate results from those interventions of life and since climate governs the kind of environment in which life finds itself, there is an intricate interrelationship between the two. In essence, life is embedded in the dynamic system which is climate.

Life on Mars?

The question has arisen whether Mars once had life on it, whether it still has, and whether life might have arrived on Earth on the back of a comet. On the face of it Mars, with surface temperatures of -53°C and a thin atmosphere mostly of carbon dioxide, would not seem a particularly hospitable place for life. All probes to date, especially NASA's unmanned, twin Viking missions in the mid 1970s, have failed to find any sign. But perhaps no one has looked in the right place?

One way to look for life on Mars is to send a Viking mission probe in the hope it might capture some living organism. Another is to look at its atmosphere. The momentous idea of Gaia struck James Lovelock with incredible clarity more than a quarter of a century ago when he realized that the fingerprint of life on Earth was there for all to see in the atmosphere, and much easier to detect than searching among the debris of sediments or waiting for a chance tiny speck to come crawling by on the surface.

Paradoxically, it was NASA's search for life on Mars that caused James Lovelock to realize how remarkably different the Earth's present atmosphere was from its two flanking planets, Venus and Mars, each of which share a similar origin to the Earth. With regard to life, that difference was crucial. Venus has an atmosphere 90 times heavier than ours made up of 98 per cent carbon dioxide, 1.9 per cent nitrogen, a trace of oxygen and an average surface temperature of 477°C. On account of its -53°C surface temperature, Mars has an extremely thin atmosphere that has a pressure more than 150 times lower than that of the Earth. Again, carbon dioxide dominates, making up 95 per cent of the atmosphere. Nitrogen comprises 2.7 per cent and oxygen 0.13 per cent.

Without life the Earth would have an atmosphere comparable to Venus, with 98 per cent carbon dioxide, 1.9 per cent nitrogen, a trace of oxygen, an atmospheric pressure 60 times that of the living Earth and an average surface temperature between 240° and 340°C.

However, with life, Earth's atmosphere is transformed, with 0.035 per cent carbon dioxide, 79 per cent nitrogen, 21 per cent oxygen, 1.7 ppmv of methane, and an average surface temperature of 13°C, and ten million times more water.

PLANETARY ATMOSPHERES				
	Venus	Earth with Life	Mars	Earth without Life
Carbon Dioxide (%)	>90	0.035	>80	98
Nitrogen (%)	1.9	79	2.7	1.9
Oxygen (%)	trace	21	<0.13	trace
Methane (%)	none	0.003	none	none
Water (m*)	0.003	3000	0.00001	0.003
Pressure (atm)	90	1	0.007	60
Observed surface temperature (°C)	477°	15°	-47°	unknown
Surface temperature in absence of greenhouse gases (°C)	-46°	-18°	-57°	unknown
Warming due to green-house effect (°C)	523°	33°	10°	240°-340°

* Depth of water in metres over the planet if all water vapour precipitated out of the atmosphere.

Figure 19. A comparison of the planetary atmospheres of Mars, Venus and Earth.

The Presence of Life

In contrast to Venus and Mars, the gases in the Earth's atmosphere are a combustible mixture, with methane and oxygen present at the same time. Given the physical and chemical conditions, and in the presence of sunlight, the Earth has an atmosphere that is far from stable, and this is the crux of Gaia theory — the extraordinary insta-bility of the Earth's atmosphere is generated by a thin veneer of life spread across the surface of the planet, which interacts energetically with its immediate environment. Moreover, as physiological condi-tions within our bodies provide a satisfactory internal environment for the functioning of tissues and organs, so Lovelock's Gaia theory maintains that when all life's interactions are combined, the result is the emergence of a self-regulating system that helps to maintain con-ditions which are suitable for life itself. Our planet's atmosphere and climate are therefore the result of this Gaian web of interactions.

Life's Ability to Transform the Environment

Twenty years ago, biologists discovered organisms in the depths of the oceans, which lived next to hydrothermal vents in mid-ocean ridges where water spews out at temperatures of 350°C. In such extreme conditions we would expect these organisms to be heat-loving bacteria, and that is just what we find. However, the surprise is that these specialist bacteria actually live in the gills and body tissues of large invertebrates, such as giant clams, mussels, snails and pogonophoran tube worms.

Using hydrogen sulphide and carbon dioxide vented in the hydrothermal stream the bacteria generate simple sugars, methane and sulphate. In return, the animals greatly extend the area in which the bacteria can live. The relationship between the bacteria and their hosts is therefore symbiotic.

From a wider perspective, the impact of the bacteria and their animal companions is vital, since it is by their efforts that sulphur, emitted from the volcanic vents as sulphide, is converted into sulphate. As we have seen, surface-living plankton, such as the minute coccolithophores, then use that sulphate to make their cell-protecting osmolyte which decomposes into the DMS that generates marine clouds. The clouds in turn carry the sulphur as acid rain, to the continents where it helps to weather rocks and soil so that nutrients, including phosphates, flush into the oceans. In time, the sulphate is drawn back into the mantle, to await its turn for fiery volcanic eruption. The sulphur cycle, is therefore driven on the one hand by geophysical processes, and on the other by life.

Living Atmosphere

Our atmosphere has undoubtedly changed considerably over the Earth's history and most climatologists, Lovelock included, are convinced that life played a big hand in transforming it to its present state. The atmosphere, with 21 per cent oxygen, is now oxidizing whereas in the Hadean and Archean periods, with no free oxygen to speak of, it was much less reactive. Radioactivity was also more than ten times greater than today and the Hadean era was consequently a time of fire and volcanoes, filling the atmosphere with carbon dioxide. The vigorous conditions on the planet also led to the generation of hydrogen, from the reaction of the ocean waters with basalt bed rock. Some of that hydrogen probably escaped completely into space. It is conceivable that had such a process continued, the Earth would have lost much of its water, since oxygen would have had a dwindling source of hydrogen with which to react.

Water: the Key

Judging by channels etched into its surface, Mars once had free-flowing water, but the water has apparently gone, although some believe there is water frozen beneath the surface. The surface of Mars, 3.5 billion years ago, was pockmarked with craters created by planetary debris which churned the surface of Mars to a depth of at least 2 km. Weathering occurred and hydrogen bubbled off from the interaction between acidified water and ferrous iron and sulphides, leaving the surface of Mars a rusty red as we see it now. Free hydrogen would rapidly have escaped from Mars' thin atmosphere, leaving the planet practically waterless.

How did Earth escape that arid fate? The critical time span for Earth when sufficient hydrogen could have escaped, leaving the planet waterless, coincided with bacteria, especially blue-green cyanobacteria, discovering how to use sunlight to make sugars out of carbon dioxide and water. Photosynthesis releases oxygen which, as well as reacting with reduced substances such as ferrous iron and sulphur in the basalt rocks, would quickly react with any free hydrogen and bind it to the Earth's surface as water. The supposition is that Earth escaped Mars' waterless fate through life's timely intervention in setting up a production line in the photosynthetic release of oxygen.

One of the remarkable findings of recent years is the extraordinary uniformity in the form and function of *chlorophyll*, which stretches from cyanobacteria, now associated with some of the first forms of life on Earth, to dramatic forms as redwoods and tropical hardwoods. All this suggests continuity and perpetual presence over the aeons. We owe our existence to chlorophyll, not simply because of photosynthesis and the capture of solar energy, but also because as a fundamental element of photosynthesis, chlorophyll made a profound difference between the Earth having and retaining water.

Had oxygen been present in the Earth's early atmosphere, life would not have come into existence since oxygen, by way of its aggressive chemistry, would have immediately destroyed the first tentative steps towards self-regulating, self-replicating organisms. The temperature and chemistry in the immediate environment of the first self-replicating entity must have been just right, which points to an oxygen-free environment. Yet, life would have been limited in its domain had it not discovered how to obtain energy from sunlight. If Lovelock is correct in seeing life as a critical component of planetary self-regulation, then without the breakthrough of photosynthesis, life would never have been able to colonize much of the Earth's surface and Gaia, as a phenomenon, would not have come properly into being.

The Toxicity of Oxygen

The irony of the Earth surviving as a *living* planet is that oxygen is a poison. The methanogens and nitrogen fixers, such as *rhizobacter* and *azotobacter*, that live respectively in the nodules of legumes such as the pea and bean, or are free-living in the soil, are unable to tolerate oxygen. One way around the problem is for the nitrogen fixers to respire frantically so as to quickly transform any free oxygen to carbon dioxide; another is to make a membrane that can keep oxygen out. Legumes have a third strategy: they synthesize a special haemoglobin *(leghaemoglobin)* that has a propensity for mopping up free oxygen in the nodules.

Nonetheless oxygen is the breath of life for creatures such as ourselves, and especially for energetic organisms such as large flying insects and birds that have high metabolisms and need to burn up sugars and fats at high rates. Organisms that live in an oxygen-rich environment have therefore developed a battery of special enzymes in their cells and tissues that quickly defuse the highly reactive radicals, such as hydroxyl and hydrogen peroxide, that are the consequences of relying on free oxygen for respiration. Vitamin E is one such antioxidant.

Evidence of Early Photosynthesis

The Barberton Mountains running from the north-east of South Africa across Swaziland to Zimbabwe, contain a mixture of unusual magnesium-rich lavas called *komatiites* and rocks called *stromatolites*, so named because of their multilayered appearance. What makes these komatiites exceptional is their age: all are more than 2.5 billion years old, and have come from volcanic lavas emerging at temperatures of as much as 1,600°C.

The komatiites were particularly prone to weathering and the oceans became increasingly alkaline as a consequence of the run-off of bicarbonates. Calcium in solution is extremely toxic to living cells, yet early life was able to survive in the oceans because the high alkalinity conferred by the weathering of komatiites caused calcium to precipitate out as its carbonate.

Stromatalites

The stromatolites also revealed a remarkable association with life. By using a diamond-tipped saw to make ultra-thin slices of hard rock from the ancient Gunflint rocks of the Great Lakes of North America, Elso Barghoorn and his colleague, Stanley Tyler, from Harvard University, found the fossilized forms of filamentous, and the more spherical coccoid cyanobacteria embedded in the fabric of the rock. The discovery took everyone by surprise; not only did it indicate that life was an early comer to Earth, but that more than 2 billion years ago it was already sophisticated. More remarkably, living descendants of similar shape and form are still to be found in various tropical parts of the planet, such as the salt lagoons of Baja California and Western Australia's Shark Bay.

Observations of contemporary cyanobacteria show that they live in a community of different single-celled organisms which lay down layers of calcified rock around them. These are later infiltrated by form-preserving silica. As the communities grow and keep pace with changes in sea level they form *stromatolite mats,* which can grow into dome-shaped structures as high as 2m and 4-5m long. One dome may finish up perched on top of another, thus forming a rock many metres high, rather like the way a well-established coral reef stands on the shoulders of layers laid down by ancient ancestors. Hence, bacterial mats were forming and precipitating out calcium and magnesium carbonates in a way comparable to today's coral reef builders, but billions of years earlier. Microbial mat communities are highly structured, with the top layers capable of coping with oxygen and the lower layers, buried in sediment, needing an oxygen-free environment.

Free Oxygen

The first atoms of oxygen released by these bacteria would quickly have been mopped up by the iron and sulphur in rocks on the Earth's surface, as well as by any hydrogen that was released by the

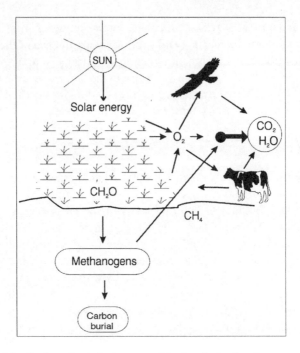

Figure 20. Carbon burial plays a critical role in the decomposition of carbon and the release of oxygen into the atmosphere. (Source: Lovelock, Healing Gaia, *Harmony 1991.)*

interaction of water and basalt lava. After more than a billion years of active photosynthesis, free oxygen began to build up in the atmosphere. About 2.5 billion years ago, at the end of the Archean period and the transition to what is known as the Proterozoic, the atmosphere changed from being a *reducing* one to one which is *oxidizing*. At the transition oxygen was still a trace gas.

Functions of Gaia

The microbial mat communities carried out a number of functions, all related to their survival: one was to speed up the precipitation of calcium carbonate thus controlling calcium toxicity: another, was to draw down carbon dioxide for photosynthesis, with oxygen released as a by-product of sugar production; a third function was to encrust evaporated salt with a gelatinous coat that prevented its re-dissolving with the next tide. Over long periods enough calcium carbonate could be deposited to form a barrier that cut off the lagoon from the rest of the ocean. Under a strong tropical Sun the water gradually evaporated, leaving crystalline salt behind. Plate tectonics would then contrive to remove that salt for geological periods out of the oceans and even secure from groundwater.

Life cannot be charged with controlling the rate of subduction of sea water, nevertheless it exerts some control over the rate at which salts are deposited in the ocean and their rate of removal. First, the release of oxygen into the atmosphere increases the rate of weathering, which brings essential nutrients, such as phosphorus, into the ocean. Second, calcium also runs off as a result of rock weathering. The process is therefore self-serving: the increase in nutrients stimulates the activity of marine micro-organisms, which then speed up the deposition of calcium carbonate.

Could the build-up of limestone reefs near the continental margins have triggered the movement of the Earth's crust that we now associate with continental drift and plate tectonics? Neither Mars nor Venus appear to have plate tectonics, although they clearly have volcanic activity and crustal upwelling. The evidence so far is that any movement of the crust on Venus and Mars is random, irregular and not conducive to the massive recycling of chemical elements that is the hallmark of our own geological system. Don Anderson, Professor of Geology at Caltech, suggests that, through the agency of life, sufficient limestone has been deposited on the ocean floor to alter the chemical composition of ocean crust at the continental margins, which therefore became far more fluid and mobile, hence moving in an organized and coherent fashion.

Oxygen Levels

The oxygen released by photosynthesis comes from the splitting of water, leaving hydrogen free to reduce carbon dioxide to carbohydrates. For oxygen to build up in the atmosphere to today's value of 21 per cent carbon needs to be buried, so that it is out of the way and not accessible for biological or physical decomposition. When organisms perish their tissues for the most part are consumed by other organisms. Putrefaction and decay are sped on their way by a whole range of consuming organisms. In the end, virtually all the primary products of the photosynthesizers are digested and returned to their original constituents. For life to exist and evolve this recycling is essential, otherwise basic nutrients would be unobtainable. Nevertheless, about 0.1 per cent of the carbohydrates is effectively put out of circulation, and it is that proportion, accumulated over time, that enabled oxygen to build up in the atmosphere to its current levels.

James Lovelock claims that the rate of carbon burial has been constant throughout Earth's history. Billions of years ago, before free oxygen had appeared in the atmosphere, the products of photosynthesis would have been largely consumed by methanogenic and other anaerobic bacteria. As oxygen levels rose and the land surface was colonized by large forms of life, such as modern vegetation including tall trees, most of the products of photosynthesis began to be consumed by oxygen-requiring organisms. In our contemporary world about 97.5 per cent of the products of photosynthesis are consumed by oxygen lovers, the oxic consumers. That leaves about 2.5 per cent for the microbes in swamps and in the guts of animals and a tiny 0.1 per cent for burial.

Once again it appears that a balance has been achieved between the burial of organic carbon and its exposure to the processes of decay and regeneration so that it becomes available again in the cycle of the elements. Certainly bacteria in ocean sediments are an important component of this release of organic carbon. They actually consume some of the buried carbon when, in the absence of oxygen, they convert sulphates in the sea water to sulphides, which as iron pyrite is buried with shales and limestone. The net result is the same — the release of free oxygen and the metabolizing of the carbon. Given enough time, some of the buried carbon forms what we now know as fossil fuels. Overall about one hundred million tonnes of carbon goes out of circulation each year, which means that 266 million tonnes of free oxygen escapes into the air.

Although carbon is being buried and oxygen is entering the atmosphere, the actual concentration of oxygen is not increasing. One reason is that new crust material is constantly appearing on the surface through volcanic activity and ocean floor-spreading, as well as through uplifting caused by plate collisions. The new basalt and uplifted granite are full of minerals that react with oxygen. More oxygen in the atmosphere means less carbon dioxide, hence less greenhouse gas, and vice versa. Having finally reached its current concentrations in the atmosphere after billions of years of slow accumulation, oxygen levels appear to be holding steady and have done so for hundreds of millions of years.

Oxygen Regulation

The proportion of oxygen in the atmosphere suits active, energetic organisms. Below fifteen per cent nothing would burn and metabolism based on the consumption of sugars would slow down. However, above 25 per cent even humid tropical forests would catch alight. Fire may indeed be one of the mechanisms by which oxygen is regulated. Yet, before we assume that fire takes oxygen out of the atmosphere we need to know a little about the substances involved. Woody plants get their sturdiness from *lignin* which they have in their cell walls to confer structural strength. The more lignin, the harder and more dense the wood, such as oak.

Trees manufacture lignin from the interaction of the hydroxyl radical derived from the splitting of water and a phenolic substance, such as coniferyl alcohol. The resulting polymer is strongly resistant to decay and even to complete burning. In the presence of fire, but a relatively poor air supply, lignin tends to convert to charcoal which may remain unchanged for millennia. Lovelock suggests that woody plants may have succeeded in making a useful structural material by detoxifying the dangerously reactive hydroxyl radical.

Vital Fires

When a tree, such as an oak, burns incompletely and charcoal is formed, the burial of carbon results. Perversely, even though oxygen is initially consumed in the fire and carbon dioxide released, this carbon burial results in a slight though significant long-term lowering of carbon dioxide in the atmosphere and conversely, a slight rise in oxygen. However, not all trees burn in this way: a resinous piece of pinewood, or a piece of eucalyptus, will practically explode into flames, leaving next to nothing but the gases of combustion. The brushwood of the Mediterranean maquis, and the chaparral in the semi-arid zones of the United States, produce volatile, inflammable oils which, by burning furiously, have the desired effect of holding back incursions of hardwood species, like the evergreen oak. At the same time, the seeds of maquis and chaparral brushwood need a superficial roasting in order to germinate. The same fire therefore clears the ground for the fire resistant seeds to get off to a quick start.

From an evolutionary point of view, it seems that the species that require fire for propagation do best when oxygen levels are on the high side of 21 per cent, since fires tend to favour them at the expense of oaks and other hardwoods. In high oxygen concentrations these latter species tend to burn to destruction. The encroachment of the brushwood leads to more fires and more advances until oxygen levels fall. Then, the fires generate less heat and hardwood species once again make territorial gains. The quantity of lignified wood therefore increases, and carbon is captured, leaving more free oxygen. Gradually oxygen levels rise again and fires kill the hardwoods and the cycle begins anew.

Oxygen and the Oceans

Not only are fires more intense and more frequent when oxygen levels are high, but more weathering takes place. Together, fires and weathering lead to more nutrients flushing out of rocks and soils, and into the oceans. The availability of nutrients in the oceans' surface layers affects the growth of phytoplankton, which in turn impinges upon photosynthesis. Many animals, when adult, lead a sedentary or even sessile life, such as coral, but during their juvenile and larval phase, they are voracious eaters of phytoplankton. The faecal pellets of these grazing animals supply bacteria, and some of the nutrients are re-circulated to the surface, through turbulence and storms, so that the cycle is still maintained. Inevitably some organic carbon escapes out of the cycle and sinks into the sediments, where it remains buried.

Theoretically, a balance of sorts may emerge out of this activity, with phytoplankton emitting just the right amount of oxygen to make up for that consumed by respiring predatory organisms. Also, organic carbon burial, may make up for the extra nutrients entering the system from a surge in continental weathering and the outbreak of fires. However, such a balance would be haphazard and easily deranged if an increased flow of nutrients were to stimulate phytoplankton activity.

Bacteria and Nutrient Control

Phosphorus is a key nutrient for life whose very scarcity has yielded surprising clues as to how oxygen levels can be regulated. When oxygen is present in sea water, any phosphorus in solution forms a complex with iron that makes it insoluble. The ferric iron/phosphate hydroxide sinks to the sediment, where it remains until conditions change. That process is helped on its way by aerobic bacteria which extract energy and nutrients when oxygen is available. The amount of oxygen in the ocean therefore determines how much phosphate remains in solution. Should oxygen levels increase, then more phosphorus goes out of solution, thereby starving surface-dwelling phytoplankton of an important resource. As a result of diminished photosynthesis, oxygen production declines and the oceans gradually emit oxygen into the atmosphere.

In oxygen-depleted waters, bacteria obtain energy for survival by reducing sulphate to sulphide. Ferrous iron reacts strongly with sulphide to form pyrite, and if it is bound to phosphate, it then releases it back into solution. The surface plankton now have access to phosphate and oxygen production can again proceed.

Philippe van Cappellen of the Georgia Institute of Technology in Atlanta, with Ellery Ingall, of the University of Texas, have modelled the oxygen-phosphorus system, taking into account the influx of phosphorus from weathering, how much ocean mixing occurs and how much sulphur and iron are available in the system to act as carriers between the oxic and anoxic states. Their model can follow the dynamics of a geological event, such as tectonic uplift of a mountain range the size of the Himalayas. Increased weathering inevitably follows and, without a compensating mechanism — in this instance, life's metabolism —would lead to a plummeting of oxygen levels within 36 million years. Life is therefore crucial in regulating oxygen levels in the atmsophere.

Continental Life

One of the greatest puzzles of the evolutionary history of living organisms is why it took more than 3 billion years before the continents were populated with multicellular life. 670 million years ago in the late Pre-Cambrian period, life took off in a remarkable and exuberant fashion. Instead of living organisms limited to single cells, individuals now became multicellular. The first of the multicellular organisms were made up of largely similar cells; then, more complex forms evolved in which cells took on different functions and became specialized tissues and organs, like livers, kidneys, brains, intestines, and heart.

The fossil record is incomplete, but one wonderful site was discovered in the Ediacara Range of hills, about 600 km north of Adelaide in southern Australia. Imprints of jellyfish, worms and a host of other soft-skinned animals have been found in the quartz rocks. The discovery of such a rich source of fossils from that period indicates that predators had probably not yet evolved and decomposition was slow enough to allow a sand cast to form.

By the Cambrian period, the proliferation of different forms was at its zenith. Every imaginable creature was to be found in the oceans, some delicate, some monstrous and rapacious. One remarkable feature was the emergence of skeletons, whether forming a hard outer shield, like the beautiful whorled shells of molluscs, or the internal calcium phosphate skeletons of vertebrates. Why the sudden use of calcium?

Some possible answers may emerge from a study of the geological processes, including rock weathering, that have taken place since the beginning of the Earth's history. Magnesium-rich komatiite lavas flowed early on, mostly if not entirely before 2.5 billion years ago, and the weathering of silicate in feldspar led to the accumulation of both carbonates and bicarbonates in the oceans, together with the minerals, sodium, potassium, calcium and magnesium. Bicarbonates confer alkalinity and under alkaline conditions calcium and magnesium carbonates tend to precipitate out as limestone and dolomite, leaving a predominance of sodium and potassium carbonate salts in the water.

We can get some idea of the rate of weathering at that time from a study of Lake Taupo in New Zealand. Two thousand years ago the central part of New Zealand's North Island blew out in a massive explosion to form a crater hundreds of square kilometres in size. Today, the amount of bicarbonate leaching into Lake Taupo amounts to 30 tonnes of carbon per square kilometre a year. At that rate 100,000 years would be enough for the Earth's oceans to become increasingly alkaline, and historically that appears to have happened.

Soda Oceans

According to geologists, Stefan Kempe and Jozef Kazmierczak, respectively from paleogeological institutes in Germany and Poland, the oceans would have become increasingly rich in sodium hydroxide: hence *soda oceans*. Such oceans could form only if the rate of silicate weathering of newly formed laval crust kept pace with volcanic eruptions and the ejection of carbon dioxide into the atmosphere.

Currently, only a few modern soda lakes exist which reflect similar conditions as the past, namely a source of weathered volcanic basalt, large quantities of a weathering agent such as carbon dioxide and an enclosed body of water. Lake Van in eastern Anatolia is the largest contemporary soda lake in the world. It has no outlet and the streams that feed it derive from volcanoes associated with plate boundaries. The fresh basalt yields high volumes of sodium bicarbonate and, like a reflection of the early Earth, stromatolite cyanobacterial mats thrive in the shallow waters at the edge of the lake. Another soda lake is Lake Nyos in the Cameroons. In 1986 carbon dioxide gas escaping from the lake killed 1,700 people and livestock as far as 25 kilometres away.

Calcium Toxicity

By the late Proterozoic era, less than one billion years ago, the alkaline ocean era came to an end, marked by the appearance of marine deposits of gypsum — calcium and magnesium sulphate — which could have formed only when calcium concentrations became high enough and free oxygen was present in the atmosphere to convert sulphide to sulphate. At the same time, the oxygenation of sulphide would have contributed to the acidification of ocean waters, which would have raised the solubility of calcium and magnesium by as much as one hundred times, so encouraging more gypsum to come out of solution.

Kempe and Kazmierczak believe that the great evolutionary thrust in the early Cambrian period — which led to the colonization of the continents — resulted from organisms devising mechanisms that not only coped with the sudden increase in calcium levels, but could actually exploit calcium's properties to develop the characteristics of modern cells.

The ability of cells to cope with high calcium levels only became possible with the modern eukaryotic cell. While the bacteria have no nucleus and an ill-defined cytoplasm, the eukaryotes have a spherical nucleus which contains nearly all the genetic material of the cell, and a cytoplasm which has structure derived from a protein microskeleton and contains a number of special bodies called organelles.

Eukaryotic cells employ many different strategies to keep internal calcium concentrations at levels 10,000 times lower than outside the cell. In common with bacteria they have a sodium-calcium pump, but in addition eukaryotes manufacture special proteins that bind to calcium and have special channels in their outer membranes that can regulate the flow of calcium in and out of the cell.

Meanwhile, the eukaryote cell has discovered how to put calcium to good use. Sponges, for instance, disintegrate when calcium levels drop, because calcium bonds and glues the junctions between cells. For the same reason, some animals, such as *Hydra*, cannot grow if calcium levels fall. The eukaryote requires calcium for the control of the spindle apparatus in cell division; for the proper functioning of cilia; for muscle contraction; for the proper functioning of nerves; for cell secretions, such as hormones and digestive fluids; for the amoeboid movement of cells such as white blood cells; for cell fertilization; and finally for the making and bonding of special proteins, tubulin and actin, used in the construction of the cell's cytoskeleton.

Complex forms of life could not have come into being had cells failed to capitalize on the versatility of calcium, and had instead succumbed to its toxicity. Undoubtedly one of the most important consequences of increased calcium levels in the sea was to bring about the appearance of algae such as the coccolithophores that could make an external calcium carbonate skeleton. The deposition of calcium carbonate beyond the continental shelves by pelagic phytoplankton such as the coccolithophores, and beginning some 200 million years ago, therefore had profound implications for the movement of carbon.

Symbiosis: a Spur to Evolution

Lynn Margulis was one of the first biologists to realize the significance of Lovelock's Gaia thesis in the early 1970s, and she introduced the notion that bacteria would have been prime movers in any process of climate regulation. In the 1970s such ideas were speculative: hard evidence had still to be found. Now, though still contentious, such ideas are gaining ground, especially with the ability to model geochemical processes, such as the regulation of atmospheric oxygen and find that without life the system oscillates wildly. Margulis' *endosymbiosis theory* proposes that bacterial activity, either free-living or from the safe haven of a modern eukaryote cell, is still the basis of Gaian atmospheric regulation, possibly of plate tectonics and if so of global climate.

One of the most radical ideas to emerge in biology over the past half century is that chloroplasts and mitochondria were originally free-living bacteria that may either have initially invaded other cells or been engulfed by others. The predecessor of the mitochondrion may have been predatory, but by learning not to kill its prey, it would have discovered that its host was a ready and continuous source of carbohydrates. In return, the host would have a marvellously efficient sugar-metabolizing package. The predecessor of the chloroplast, on the other hand, was a photosynthesizing bacterium that could generate its own sugars from carbon dioxide and water. If the chloroplast's ancestor could avoid being digested, the host would no longer need to indulge in predatory behaviour.

Ironically, whereas the eukaryote cell learned to deal with high calcium levels, the mitochondrion and the chloroplast never lost their vulnerability. These organelles still maintain calcium at low levels inside their structures by using a sodium-calcium pump similar to that found in cyanobacteria. Perhaps, their original progenitors were saved by endosymbiosis, which put them into the safe environment of the eukaryote.

Undoubtedly endosymbiosis has been a major spur to evolution. The ancestors of the mitochondrion and the chloroplast would have been unlikely to conquer the land if they had remained as a single prokaryote instead of throwing in their lot with eukaryotic cells. When we enter the Amazon rainforests, or the great boreal forests of Siberia, or if we are harvesting our crops of rice or wheat or even playing tennis on a grass court we do so courtesy of the chloroplast, which had its lowly origins in the stromatolite mats of the Archean period, simply because our ancestor cells took the mitochondrion on board.

What we find on Earth is an amazing set of chemical recycling factories, managed and operated by life. Brute geophysical forces, such as plate tectonics and volcanism, as well as the Sun, are the engines which move the raw materials of the Earth's surface layers from one place to another. Life takes the potentially destructive power unleashed on Earth and transforms it into creative, discrete forms that embody efficiency and the economical use of resources. Whether we look at the carbon, sulphur or phosphorus cycle, or any other elemental cycle, their sheer breadth and scale indicate that life is a planetary phenomenon.

Chapter Nine: When Gaia Fails

WHEN James Lovelock first proposed that life on Earth might actually regulate climate and keep surface temperatures within tolerable limits he met considerable opposition, not least from biologists. Here, we have momentous forces — volcanoes, storms, earthquakes, plate tectonics, orbital changes, and fluctuations in the Sun's output — how could life possibly govern the outcome of these events, especially if it had no prior awareness of what it was doing? On the contrary, biologists argued that even if life modified its environment by altering the composition of the atmosphere, and even if such alterations might make conditions more comfortable, the onus was still on life to adapt to its environment as best it could, through the gradual changes wrought by evolution.

Biologists therefore deemed life to be a lucky break and its adaptations simply the result of beneficial mutations that happened to make survival easier. But, was it simply luck of the draw that gave life the capacity to keep pace with changes to the Earth? On the contrary, argues the biophysicist Mae-Wan Ho: for her, luck and random events had little to do with life's coming into being and its extraordinary ability to modify not only its internal structure and activities, but the external environment. If the creation of the first life forms, 4 billion years ago, were simply the coming together of the right basic constituents then, according to classical thermodynamic theory, the probability of life happening would be extremely small. The universe would neither have had enough time nor matter to bring about such an event. But life is here and has been here for considerable time, modifying both itself and its environment. In *The Rainbow and the Worm,* Mae-Wan Ho claims that the key to life's remarkable prowess is the way it structures itself, as if it were a superefficient, solid state, physical system.

As Mae-Wan Ho makes clear, by the very nature of its precise structure, life can capture the energies of electrons that have been excited into higher energy states, by intercepting a photon from the Sun. Finally, after passing through a cascade of organized molecular interactions, the electron returns to its *ground* state. Life's extraordinary property is space-time organization on an unimaginably complex scale.

Nothing exemplifies this more than photosynthesis, in which chlorophyll molecules are able to capture photons and stream them down to activation centres. One amzing finding is that the numbers of aligned chlorophyll molecules in a leaf — more than one thousand million million per square centimetre — are at the correct density to absorb most of the right wavelength photons of daylight hours.

The potential exists for considerable efficiency, although the net result is determined by other factors such as the availability of water and other nutrients. Of 240 Wm^2 received at the Earth's surface, barely 0.7 Wm^2 worth of energy gets through for the production of sulphur biomass.

However, at least one third of sunlight goes into the hydrological cycle and another third drives the convective forces which move air masses, carrying the vapour that irrigates the land. At every step energy is lost, although we should never regard it as a loss, since the energy created the very dynamic that has brought about the Earth we live in.

Superefficiency

The fundamental difference between man-made machines and life is the exquisite efficiency with which energy is employed in living processes. The eye, for example, can detect a single photon and by a series of molecular transformations can generate an electrical impulse to the brain that may be one million times more powerful than the original signal, and all in no more than one hundredth of a second. Energy in the living cell is further held and partitioned by the intricate web of different interacting forms. Primary producers — bacteria and plants — are consumed by a range of different grazers and herbivores, that themselves fall victim to predators. Finally, in death, virtually all life's forms decay and decompose, helped on their way by fungi, bacteria and even scavengers such as hyenas and vultures. The more varied and diverse the system the greater likelihood that it will be energy and resource-efficient. A degraded life system is inherently inefficient and, with its potential to alter the Earth's surface properties, including atmospheric chemistry and climate, is probably non-sustainable.

Weathering and Efficiency

In the 25 years since Lovelock first formulated his ideas on life's role in regulating surface phenomena on our planet, biologists, oceanographers, climatologists and geologists have been gathering evidence of the power of living processes to transform their environment.

Life, for instance, accelerates weathering one hundred times or more through drawing down carbon dioxide. New, exposed, continental rock is soon attacked, by bacteria, algae and fungi, often symbiotically combined as lichens, and then by other plants and a host of burrowing creatures, not least earthworms and termites. Trees, because of the large surface area of their leaves, are particularly effective in bringing carbon dioxide out of the atmosphere. Their roots penetrate deep into the soil and subsoil. But they do not operate alone, and special fungi, called mycorrhizae, live in close symbiotic association with the root systems of healthy trees, as well as other plants. The fungi accelerate weathering around them, and, helped by the mild acid solution of carbon dioxide, are able to mine the subsoil and even bedrock for minerals. They then form *bridges* through which they pass on essential nutrients, such as phosphorus, that they have made available. It is even conceivable that the mycorrhizae may create bridges between one tree and another, so that a root mat develops in the soil which is interconnected. In the washed-out, nutrient-poor soils of the *blackwater* regions of the Amazon Basin, flanking the Rio Negro, the root mats are so efficient at recycling minerals and nutrients that the rainwaters flushing into the river are almost as devoid of minerals as is distilled water. When a tree dies, the root mat soon extends over it and its decomposition leads to a nigh perfect redistribution of its original constituents.

Carbon Recycling

One of the attributes of life on Earth is its phenomenal ability to keep essential elements in circulation. In *Gaia's Body,* Tyler Volk gives us some examples of life's recycling of the six main essential elements embodied in the acronym CHNOPS — carbon, hydrogen, nitrogen, oxygen, phosphorus and sulphur. As he points out, global photosynthesis uses a total of one hundred billion tonnes of carbon a year — nearly fifteen times more than the carbon we push into the atmosphere from our consumption of fossil fuels and our destruction of forests. Photosynthesis on land captures sixty billion tonnes of the total carbon and the oceans the remaining forty billion tonnes. With a total of 750 billion tonnes of carbon in the atmosphere, were there no recycling of carbon through respiration, in less than ten years all the carbon dioxide in the atmosphere would be depleted. In fact, volcanoes and rock weathering put about half a billion tonnes a year of carbon back into the atmosphere — which coincides with the amount of organic carbon buried through natural processes.

Consequently, without the recycling, photosynthesis would have to make do with the half billion tonnes of carbon from geophysical processes. Life would be a marginal phenomenon and Gaia, as a planetary system, would simply not exist.

Nitrogen Recycling

Obviously the recycling of the other elements must keep pace, at least in proportion to their use in the creation of organic molecules. 99.6 per cent of all the nitrogen on Earth is in the atmosphere. As Tyler Volk remarks, it would take no more than half a million years for the nitrogen in the atmosphere to be used up, were life doing nothing to get it back. If that were the case, then the oceans would gradually mop up nearly all the nitrogen, and life on land would not be possible. We must thank the denitrifiers which put back the nitrogen as gas; equally, for the equation to balance life requires nitrogen fixers. It is they, tucked away anaerobically in the soil and plant nodules, that make life on land possible. Remarkably too, the concentrations of nitrogen in the atmosphere dampen down oxygen's propensity for setting everything alight. It all seems too good to be true, if it were nothing but chance. Instead, as Volk points out, we must thank bacteria for amplifying the cycling ratio of nitrogen between 500 and 1300 times.

However, we still face the question of how life could possibly know what to do to regulate climate. How could it have realized that generating clouds over the oceans or fog over stretches of *Sphagnum* dominated tundra would keep surface temperatures down, or equally that spring and summer could happen sooner if the vegetation was as dark and heat-absorbing as are the pine needles of high latitude or high mountain conifers? How could life know that the drawing down of carbon dioxide to its current levels would make the greenhouse content of the atmosphere just about right for today's solar input on to our planet?

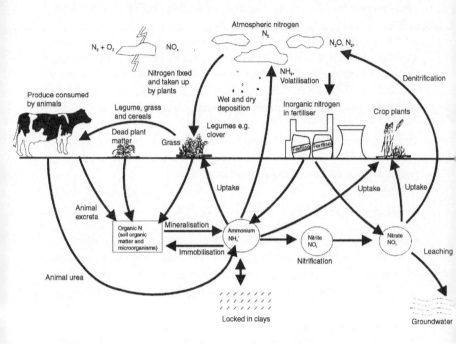

Figure 21. Nitrogen recycling.

These are formidable questions and unless answered satisfacto-
rily scientists will remain sceptical that life is anything more than a
tinkerer when it comes to the processes of planetary regulation.
Lovelock accepted that the onus was on him and his collaborators to
come up with a plausible way in which life could bring about regulation, while remaining unaware in a conscious sense of what it was
doing. For many it goes against the grain that unconscious life may
be effective in generating a workable system of regulation on a
global scale. That would diminish our claim that we know better
than Nature and that through our knowledge we can manage the system intelligently rather than leaving it to haphazard forces. Are we
not now managing the ozone hole by limiting the use of certain
chemicals, or on our way to managing greenhouse gases, by planting forests of carbon-mopping trees, like the plantations of Monter-
rey pine down the length and breadth of New Zealand's North
Island?

Daisyworld

Lovelock insisted that global management is effective primarily because it is unconscious. Together with Andrew Watson, then at the Plymouth Marine Biological Laboratories, he came up with a simple mathematical model, called *Daisyworld*, to show that life could have the ability to regulate surface temperatures on a global scale without any preconception that such would be required for life's future survival. In its first version Daisyworld receives the same amount of energy as the Earth and shares the same history of orbiting a star that becomes more luminous as it ages.

If Daisyworld had no life at all, then initially the amount of light striking the surface would be barely enough to raise the average temperature to 0°C. As the Sun heated up and emitted more light the hypothetical planet would get hotter until, 4.5 billion years later, its surfaces exposed to the Sun would be unbearably hot.

That is where the daisies come in. Lovelock and Watson conjured up two distinct species of daisy, one black and one white. The black had an albedo that made it absorb more heat from the Sun than the surrounding rock, and the white daisy had an albedo that would reflect much of the Sun's energy. Like mainstream life on Earth, the daisies would have difficulty growing at all when the temperatures were 5°C or below, and would tend to overheat and stop growing once the temperature had reached 40°C. Hence the daisies did best at an equitable temperature of around 22.5°C.

What first happens in the model, is that when the local temperature has reached 5°C the daisies of either species start growing, but not very well. At this point the black daisies tend to make their local environment a little bit warmer than it would be without them and that extra warmth encourages them to grow better, and then better still. The white daisies, on the other hand, make their local environment a little cooler and have difficulty getting off to a start. They could try and benefit from the warmer areas where black daisies are growing, but will be smothered as they make conditions for themselves less favourable. Not only will the black spread more across the planet, but in the early stages their growth and spread will accelerate through self-induced advantage, and will therefore be subject to *positive feedback*. In time, the black daisies will compete with each other for space, and their growth will taper off. Moreover, an increase in the areas with black daisies will raise the temperature above the optimum. *Negative feedback* now occurs with too few blacks leading to excessive cooling and too many to excessive heating.

Once the surface temperature reaches the optimum temperature for growth, the increasing luminosity of the Sun favours increasing incursions by white daisies into areas previously held by black. An observer ignorant of the history of the solar system and believing surface temperature to be a given, might view the spread of one daisy at the expense of the other as the result of brute competition. However, if the observer had prior knowledge of the system, he or she might surmise that the temperature was an emergent property of the total system; daisies and environment combined.

Daisyworld's Parallels with Earth

Although simplistic, the Daisyworld model offers fascinating parallels with Earth. First, the build-up of warmth on the planet, once life is there, precedes the Sun's ability to warm the planet. Equally Daisyworld retains its habitable characteristics longer than it would were life not there. Yet, in neither instance do the daisies have any idea what they are doing with regard to regulating surface temperature. The exciting characteristic of the model is its indication that life's ability to change the surface absorption of heat is enough in itself to allow the property of temperature regulation to emerge. Daisyworld is therefore the best possible world because life has made it so, and without life, Daisyworld would become inhospitable far sooner than is the case when life is present.

Has life in the main adapted to what it finds, or has it actually moulded what is there to provide an optimum environment for its own growth and development? Perhaps these questions can never be categorically answered one way or the other, but at least the models begun by Lovelock and others have challenged the orthodox view of *evolution by chance.*

Since Daisyworld, Lovelock has expanded the basic ideas of the original model and incorporated greenhouse gases, simple atmospheric chemistry and cloud cover into the equations. Through such models he has therefore looked back over the possible history of climate, incorporating what geology tells us about life's past. By putting life and certain environmental parameters into the differential equations, he finds, throughout his simulated history of the Earth, that life brings about average surface temperatures that never stray too far from the levels which support life as we know it. The dynamic interaction leads essentially to an Earth that exhibits regulation.

Can Gaia fail?

Can Daisyworld offer us any clues as to how the Earth might be faring today as a result of global warming? Lee Kump and James Lovelock set up a model specifically to look at the respective contributions of life in the oceans, and the stabilizing of global temperatures on land. The contribution from the oceans was the formation of clouds as a result of surface algae generating dimethyl-sulphide (DMS). The contribution over land was from the drawdown of carbon dioxide by vegetation.

The model operates on the principle that as the oceans warm the thermocline layer spreads and becomes a greater barrier to the mixing of deep and surface waters. On land, as temperatures increase, plants find themselves increasingly suffering from the drying out of soils and water stress. In both instances, rising temperatures take their toll on growth. At a certain point, life in the oceans and on land loses its ability to counter global warming and any vestige of regulation disappears. The model shows clearly that life in the oceans is more vulnerable to higher temperatures compared to life on land. The thermocline barrier is a feature of oceans when the surface waters are warmer than about 12°C. On land, water stress becomes a widespread phenomenon once the average surface temperature reaches 20°C.

Oceanic Thermostat

During glacial conditions both the oceanic and terrestrial systems regulate surface temperatures, with ocean algae doing best in the warmer range of the glacial period, when able to spread to lower latitudes (see Chapter 6). As long as the ocean mixing is good, then they thrive and generate DMS, which causes clouds to form, which in turn bring about cooling. As a consequence of the falling temperature, the algae generate less DMS, and therefore fewer clouds. The Sun therefore shines directly on to the ocean surface and warms it. The feedback between the algae and temperature is therefore like a thermostat and regulation is possible. Once the surface temperature exceeds 12°C then the ocean thermostat can no longer work properly. Lovelock and Kump conclude that because of global warming the Earth is now moving rapidly towards that stage.

Terrestrial Thermostat

However, we have some way to go with regard to the terrestrial thermostat which operates through vegetation drawing down carbon dioxide. Again, it works best overall during an ice age, when more land is exposed through a fall in sea level, and plants, including bacteria, are able to bring about extensive rock weathering (see Chapters 4 and 7). The drawdown of carbon dioxide through surface weathering takes about 100,000 years, which fits well with the records of low carbon dioxide concentrations in the atmosphere at the end of the ice age. The precipitous rise in atmospheric carbon levels at the end of an ice age may result from sudden. sharp global warming, caused by rotting vegetation releasing considerable quantities of methane, bringing about a rise in sea level, and thus the flooding of coastal plains.

Temperature Regulation and Carbon Dioxide Levels

Carbon dioxide levels in the atmosphere are currently around 350 ppmv. In Lovelock and Kump's model, oceanic and terrestrial regulation operate best when the concentrations are approximately 200 ppmv. When they reach 400 ppmv the algal system collapses and surface temperatures rise quickly to another equilibrium regulated by terrestrial life. If concentrations rise to 700 ppmv then terrestrial regulation collapses. At that stage we will be in uncharted territory.

Without doubt the model is simple and cannot reflect the sheer complexity of the real world, but it is disturbing that we are now within a century of entering that zone where the model indicates that planetary regulation will collapse. It may be that the idea of Gaia is fanciful and that no such system operates on this planet and that the system operates purely through geophysical forces. But if we are wrong about this — and our industrial society behaved as if life were not part of a regulatory system — then we may already be in grave trouble.

Inadequacy of Models

None of the General Circulation Models of climate (GCMs) used by the IPCC as yet incorporate the notion of life as a regulator of climate into their workings. Consequently, the modellers and those they advise simply have no idea whether life is capable of reducing global warming, whether it is actually doing so, or whether its capacity to do so has already been surpassed. If the idea of Gaian stability is mistaken and life has no hand in climate regulation, then GCMs might indicate what is in store for us with regard to global warming. But, what if the Gaia thesis is essentially correct, and the global climate is still subject to some biological control? What would that tell us about the future?

Certainly the possibility of self-regulating feedbacks has not been built into GCMs, and it would be disturbing if Lovelock and Kump are mostly right that temperature regulation collapses once carbon dioxide levels exceed a level, such as 700 ppmv. Consequently, global warming could be far worse than anticipated by the most pessimistic GCM. Thus, ecosystems might be disrupted on a massive scale, threatening life as we know it.

At this stage we can only speculate on the likely outcome of continued global warming and scientists should caution that the results of their models are based on an optimistic and probably unrealistic view of the fundamental processes that underlie climate. It is essential that Gaian feedbacks on cloud formation and greenhouse gas accumulations are built into GCMs to see whether substantial differences arise between these results and the more orthodox model results that are currently in circulation.

Chapter Ten: Culture and Climate

Humans, *Homo sapiens,* emerged 100,000 years ago from early hominids that lived in Africa between 4 and 5 million years ago. We lived as hunter-gatherers, shifting in small bands to pursue game, or conceivably to scavenge what was left by other predators, and to follow the harvests of fruits, berries and tubers as they came into season.

What brought about our evolution? In *Sixth Extinction: Biodiversity and its Survival* Richard Leakey and Roger Lewin propose that a dramatic change in climate in East Africa, caused by the tectonics of the Rift Valley, was a probable catalyst. New mountains arose on either side of the rift causing a rain shadow that put paid to the tropical forests where the pre-hominid apes had their domain. To survive the pre-hominid creature had to learn to live among the thorn bushes of the savannah, and to compete for food with sharp-clawed, ferocious predators. Wits and the ability to make tools and weapons became prerequisites for survival.

In his book, *The Day before Yesterday: Five Million Years of Human History,* Colin Tudge speculates on the opportunism of our hominid ancestors in moving first from dense forest to woodland and then out into the open. He and others suggest that the essential moment for hominid evolution resulted from a dramatic chilling and drying out of climate 20 million years ago. Forests in the Tropics gave way to grassland, opening up a new habitat that an intelligent, weapon-bearing bipedal hominid could use to advantage.

Tudge pays homage to Maureen Raymo and her colleagues at MIT for their idea that it was the rise of the Himalayas that gradually depleted the atmosphere of its most important greenhouse gas through weathering. The *Australopithecines* made it to man, because they retained many basic tetrapod characteristics which they then put to good use in moving out of the forest to the grasslands which, with the change in climate, had been expanding across the globe.

Tudge views climate as a fundamental force in driving evolution, paradoxically by bringing about extinction. Hence, a dead-end for one organism, because it has encountered its climatic limits, becomes the stuff of life for another, which in the Darwinian sense happens to be that little bit better adapted, or just lucky — like the *Therapsid* proto-mammal that survived the dinosaur extinction.

The list of extinctions is staggering and climate change would appear to be a secondary, rather than the direct cause, in that it brought about the evolution of humans. Agriculture, Tudge points out, was the ultimate weapon, since ecosystems could be subjugated to satisfy human needs, while fencing out or wiping out animals that refused domestication (including other humans!). Africa proved the exception, as some have suggested, because wide tracts were protected from human invasion by the tsetse fly and sleeping sickness.

Not all mammals fared badly at the hands of Man. The hangers-on, not least the pests, found human settlements a rich source of pickings, while animals that proved amenable to domestication such as cattle and other ruminants unwittingly assisted Man's conquest of the continents through their conversion of the undigestible cellulose of grassland into rich protein, whether as flesh, blood or lactose. Grassland and savannah is therefore the key to human evolutionary success and it is perhaps not surprising that so much of the world has been converted into pasture and cattle culture.

Mass Extinctions

Species have come and gone, and the world has suffered at least five major extinctions since the Cambrian period, 570 million years ago. One hundred million years later, at the time of the geological period known as the Ordovician, at least two thirds of known animal phyla — the branches of different species — vanished. Luckily for us, one survivor was *Pikaia gracilens,* a type of worm with the beginnings of a backbone. Further evolution turned the descendents of that lowly creature into the first vertebrates.

These extinctions were probably caused by an asteroid smashing into the Earth, as were four subsequent mass extinctions. The most drastic appears to have occurred more than 200 million years ago, when ninety per cent of all species disappeared. The most recent mass extinction was that which wiped out the dinosaurs 65 million years ago.

If we look back over the history of mankind, humans seem to have wrought destruction wherever they have gone. Leakey and Lewin claim that our species has had the impact of a large asteroid in bringing about mass extinction. But whereas nature heals itself after a calamity such as an asteroid, our tendency is to keep forcing change, rarely allowing a new balance to come into being. In recent years we have unleashed a barrage of change on the environment, destroying any vestige of stability.

By the Enlightenment, scientists such as Count Buffon in France came to the conclusion that life was part of a process of evolution rather than having been created *de novo* by God in a matter of six days. Fossils were now being discovered, and by seeking an explanation, the nineteenth century French naturalist, Baron Cuvier, came up with the view that, at various times through its history, the Earth had been wracked by cataclysms which destroyed entire species. Life then had to pick itself up and start again, but based on the fundamental forms that had preceded the event. His views were prescient, and we are now finding increasing evidence throughout the geological record of mass extinctions.

Adaptation and Survival

In the long run, adaptation to the local environment, including its climate, is a prerequisite of survival. Animals and plants therefore exhibit an amazing repertoire of adaptations in their ability to cope with markedly different situations. The differences lie in the way certain attributes become prominent, while others take a back seat or even vanish. Marine mammals have turned their digited limbs into fins, while bats have turned theirs into wings. Nor is it just the external form that adapts, but also the nervous system, with specialist priorities such as an emphasis on vision, sound, touch, or smell.

Physiology expresses a particular way of life. The polar bear can tolerate concentrations of vitamin D that would be toxic for other mammals. The Koala bear and the Amazon sloth, have developed metabolisms which can cope with toxic chemicals in the leaves of their favourite plants. The fennec fox has oversized ears not just for picking up the slightest sound, but also for cooling, while Grant's gazelle can survive even when its body temperature climbs as high as 46°C.

It is the role of the ecologist to establish what the special adaptations are that give one organism the edge over another or enable it to live in close relationship with others. Since climate is one of the prime factors that determine an environment, ecology becomes the study of adaptation to a particular microclimate.

The Cosmos and Order

Cosmologies are a system of beliefs that try to make sense of the world in which we live. Tribal peoples throughout the world came to see everything in the natural world fitting into a universal pattern. Ritual, emerging out of a deep respect for the forces of nature, was a way to reiterate cosmic order and to imbue the group with a sense of unity and purpose.

Edward Goldsmith espouses the notion that traditional culture is bound to a prerogative to maintain harmonious balance in the environment. Culture, therefore, serves a function, and only those cultures that enable a responsible interaction with the environment have any long-term chance of survival.

Determinism and Culture

Anthropologists have long been involved in a debate as to how much cultures have been influenced by environmental factors — whether climate, for instance, has moulded and even determined strategies for survival. At the same time, they have become increasingly sceptical that indigenous cultures are imbued with a conservation ethic. Yet, contemporary anthropological studies give no indication whether shamanistic restraints on environmental abuse — for instance, hunting to excess — did apply in the past, before such cultures had encountered the persuasive powers of consumer-oriented cultures.

Life in Tropical Forests

Humans have mastered ways of surviving in tropical rainforests, even when soils are extremely poor. Survival depends on traditional wisdom and knowledge passed down through the generations. The Amazon, like all tropical rainforests, is made up of various environments, including dry land, flood plain, swamps, lakes and rivers, as well as patches of forest that have been cleared for gardens and the planting of fruit trees. A typical family group living in the north-west Amazon of Colombia will subsist from an area of jungle containing at least fourteen clearly defined habitats, and will harvest timber and products for their own use from 170 wild tree species, and twelve species of cultivated fruits. Hunting and fishing provide most of the protein.

The traditional economy of Amazon peoples is the antithesis of the market economy whereby individuals seek self-enrichment, often at the expense of others. Contemporary anthropologists tend to disregard the notion that a sense of reciprocity prevails in the dealings of tribal peoples with their environment and, on the simple grounds of hunting behaviour, have called into question the notion that indigenous peoples of the Amazon are conservation-minded by culture.

If hunting behaviour was linked to the maintaince of an ecological balance, under the watchful eye of a shaman, a hunting group would avoid killing pregnant or lactating female animals. The application of a statistical approach suggests that no such discrimination takes place. Hunters kill whatever they can get away with. However, such studies are based on a contemporary situation in which goods are sold rather than exchanged, and human behaviour is increasingly individualistic rather than communal.

Colonization

Our success in colonizing the planet comes from our ability to manipulate our environment and isolate ourselves as far as possible from the extremes of climate. We have therefore learnt to build houses and other structures that reflect the climate in which we live. Alpine houses have roofs that shed snow; Himalayan houses, use sturdy timber uprights and intersections to make the buildings earthquake-proof. In swampy, flood-prone areas, people build their houses on stilts, or if subject to tropical cyclones, use thatched structures which, if they collapse and blow away, will not kill the inmates.

Humans, like other organisms, can adjust to different climatic and environmental conditions, by a combination of long-term evolutionary adaptation with short-term adjustments. Adaptation involves genetic change and the supposition that any advantageous genetic mutations will be selected. Adjustment may involve physiological changes as the organism develops, and so determines permanent responses to environmental conditions throughout the life of the individual. Adjustment may be a quick and a non-permanent response to the environment, such as the thickening of the blood and the build-up of oxygen-carrying haemoglobin when people from the plains climb at high altitudes.

The anthropologist, Frank Boas, suggested that while nature imposed constraints on human culture it did not determine culture. Humans therefore had choice in what they selected from nature. In the Arctic, for instance, the Inuit subsist on hunting and fishing, whereas in a similar environment in Siberia, the Chukchee survive by breeding reindeer, as do the Sami people of Lapland. Equally, in the semi-arid terrain of Namibia and South Africa, the Hottentots lead a pastoral existence, while the Bushmen rely on hunting and gathering. Culture is therefore a dynamic entity which actually shapes how the environment is exploited and lived in. A similar environment may thus present entirely different solutions.

Physiological Adaptation

The Inuit live in an extreme environment, and not only face bitter cold in the winter, but also months without light or sunshine. Their adjustment to climate involves both culture and physiology. The biological productivity of the tundra is low, and the Inuit discovered long ago that more was to be gained by hunting marine mammals rather than relying on the migrations of caribou and musk ox. Like Amazonian peoples, who traditionally view the forest as inhabited by guardian spirits that protect life and therefore the hunted, the Inuit have a healthy respect for their prey, seeing them as imbued with spirits which, when freed by the kill, return to the ocean to find another body to inhabit. Over-hunting is thus an abuse which results ultimately in catastrophe.

We usually think of shivering as the way the body tries to keep warm. However, the Inuit can raise their body heat without involving any muscle movement by increasing the rate of metabolism thirty or forty per cent more than people of more temperate climes. Inuit adults retain *brown adipose tissue* which is normally found only in infants; this fat provides the additional metabolic warmth. The Inuit are also able to quickly send blood to the extremities should they be exposed to the cold. Such blood vessels have also been found to dilate and constrict in a cyclical fashion to prevent frostbite.

If he is to survive, an Inuit hunter of the Canadian Northern Territories has to recognize weather conditions that give rise to blizzards. His survival then depends on his skill and dexterity in erecting an igloo out of ice, before the landscape is blotted out with snow. The right kind of ice for the igloo is full of air pockets, which therefore improve insulation. The shape too is perfect for minimizing wind resistance, while giving adequate space inside for the least surface area. Body heat plus a seal oil lamp, which also gives lighting, is sufficient to make a cosy place of refuge.

Changing World

Much has now changed through the Inuits' encounter with western technology. Instead of dog sleighs the inuit resort to snowmobiles and hunting with guns. Education has also become classroon orientated, and the skills of a hunter, acquired by a childhood spent observing adults, have declined. Obesity, diabetes, high cholesterol levels, anaemia and heart disease are rampant. The Inuit no longer live in traditional dwellings, but have moved to pre-fabricated structires heated by burning coal, resulting in an epidemic of respiratory infections.

Mountains and Culture

Mountains are also a marginal habitat for humans. As with the Inuit, no marked genetic differences have been found between the Andean people, such as the Quechua, and other human populations. However, Andean peoples, living at altitudes where atmospheric pressure is 65 per cent that at sea level, have developed physiological changes in lung capacity, in the network of capillaries in alveolar tissue, in haemoglobin-enriched blood compared with people at sea level.

Through complex family relationships and trading partnerships, the Quechua have developed a strategy whereby they can gain access to lands at different altitudes. At 4,000 metres they grow highly-nutritional chenopods, such as quinoa and cañihua, and innumerable varieties of potato. At lower altitudes, they grow pulses and tubers, By exploiting the range of habitats, microclimates and soils that come with different altitudes, the Quechua are able to spread and manage the risk of living in a marginal environment.

Variety of Adaptation

When we study all the major habitats in which humans live, including grasslands, tropical rainforests, temperate lands, as well as the tundra, altiplano and desert, we find in all instances patterns of survival that rely on immense, handed down knowledge of the environment and its medley of plants and animals. Not that there is necessarily a single solution to any one situation, and different strategies have proved to be successful in comparable environments.

However, it is remarkable how humans have hit on social systems that serve to maintain balances over such a variety of environments.

Imposed Culture: the Consequences

Modern *westernized* civilization has now taken over the best lands and remorselessly forced tribal peoples and traditonal farmers to the margins. In this respect the European plough, with its mould for creating deep furrows, was taken to Africa to replace the traditional plough which left a shallow furrow. Initially the European way led to higher yields and no one bothered to look at the irrevocable damage to soil perpetrated through deep ploughing. In the Sahel countries, for instance, the colonial governments encouraged villagers to clear their land for the plantation of cash crops such as peanuts. Incentives in the form of fertilizer subsidies were used to promote such development, and villagers were told to break with the tradition of leaving land fallow for the herds of pastoralists. In the meantime, the colonizers dug boreholes in the savannah for watering livestock and carried out various veterinary campaigns to reduce disease. The pastoralists, to compensate for the loss of access to the villages, were informed that they would now have access to markets overseas.

The modern strategy proved a disaster. The villagers found the fertility of their land fell rapidly, at the same time as subsidies for peanut production ceased. The pastoralists, meanwhile, had allowed their livestock populations to increase to the point where overgrazing became a serious issue. Then, in the late 1970s and early 1980s, the rains failed over the entire Sahel. The land was devastated and millions of cattle died. To date, the Sahel has not recovered. In a matter of years, western interference destroyed a way of life that had survived relatively untouched for thousands of years. Moreover, the destruction of vegetation altered the albedo and energy balance, instigating a process of local climate change that could lead to even more drying out.

Death of Culture

Few cultures have survived the onslaught of the industrial age, and whereas traditional cultures were sustained by natural energy and nutrient flows, modern cultures have become increasingly dependent on a flow of energy and materials that are outside natural exchanges. Instead of using technologies to adapt to the environment, we now use them to force environments to suit us. The problem is that modern technologies have transformed the world so much that we no longer see the consequences of what we are doing, deluding ourselves that the Earth is resilient enough to take what we are doing to it.

Our cavalier attitude towards the environment has undergone a rude awakening. We are shaking the planet to its roots: causing global warming; damaging the ozone layer; and disturbing the natural cycles on which we have always depended. Ironically, our use of technologies to make us independent of the natural environment, is generating changes that will ultimately force themselves upon us. Climate again will become a determining factor and, if we are to survive, we will have to learn to adapt.

Chapter Eleven: Land Use and Climate

More than forty per cent of the Earth's land productivity has now been sequestered by humans and, with the world's population set to double over the next century, greater land usage is likely. Inevitably, our planet will become impoverished as species are squeezed out and disappear. Pests and pathogens will flourish as monoculture crops take over from natural ecosystems.

The changes in land use across the globe have undoubtedly contributed to climate change, through changes in albedo as one kind of vegetation is exchanged for another. The change in the flow of water through the environment as a result of agriculture has also affected energy transfers and cloud formation, and our efforts to increase yields through the use of fertilizer-responsive crops has led to a doubling of the flow of nitrogen through the natural system.

Nitrogen is fixed industrially as nitrate and ammonia for application to crops and, together with nitrogen oxides generated in combustion engines, is now beginning to flow in much greater volumes through the planetary ecosystem, compared with pre-industrial times. Part of the nitrogen waste problem results from the growing population of livestock kept under increasingly intensive conditions. Worldwide the numbers of cattle now exceed one billion, and their impact on land in terms of the flow of nitrogen waste is considerably greater than that of humans.

Many plant species, especially trees, cannot tolerate high inputs of nitrogen and become vulnerable to disease. Cattle, especially those kept on poor grazing, also generate large quantities of methane which, like nitrous oxide, is a powerful greenhouse gas. With additions of the gas from increased rice production, the concentrations of methane in the atmosphere have been rising by more than one per cent per year.

Population and Resources

The human population has been increasing at a high rate since World War Two and more people have been born into the world and survived over the past fifty years than have ever existed before. In 1950 the world population was just over two and a half billion, with an annual addition of some thirty seven million; by 1994 the population had risen to over 5.6 billion with an annual addition of eighty eight million. In some countries, the growth rate has remained high, setting the population on an exponential path. Pakistan, for instance, is likely to become the world's third most populous country over the next fifty years, despite high infant and maternal deaths. On average, Pakistani women now have as many as 5.9 children compared to 3.4 children per woman in India and Bangladesh. Meanwhile, India's population has tripled in fifty years and is still expanding, and both its population and China's now exceed one billion. Current trends indicate that the world population will reach six billion by the turn of the century, and will double again within another fifty years, that is assuming an average rate of increase of 1.5 per cent per annum. In fact, the average population growth rate has come down from a peak of 2.2 per cent in 1963 to just over 1.5 per cent today, and the United Nations projects that the rate will decline slowly over the next few decades. However, not only do more people remain alive each year, but the population has become increasingly young, with approximately one third below the age of fifteen.

Paradoxically, the youthfulness of the population of today's developing countries means that their crude death rate may be considerably lower than that of a more highly developed country, such as the United States, with its population spread out among the different age groups. A youthful population, combined with slightly longer life expectancies, is a recipe for accelerated growth and that is what we find in many developing countries in Asia and Africa.

The inexorable dictat of matter and resources means that population growth has real limits and, should growth continue, the limits will make themselves felt within a short time, compared to human history. As the physicist Isaac Asimov, has pointed out, were the human population to increase at a rate of two per cent per annum then, in less than 2000 years the mass of humanity would exceed the mass of the Earth. In less than 6000 years, the mass of humanity would exceed the mass of the universe. Such calculations can be applied to any organism, and bacteria which have the capacity, given all the nutrients they need, to reproduce in minutes, would take days rather than thousands of years to be brimming out over the edges of the universe. However absurd such notions may be, they have the salutary effect of bringing us down to Earth and making us realize that the limits to growth may be closer to hand than we take into account in our *economic* planning for the future.

Environmental Abuse

Another complication in the relationship between resources and population size is the effect of technologies that extend the resource base, such as the widespread use of crops designed to respond to industrial fertilizers by doubling or tripling the yield of edible grains. The International Rice Institute in the Philippines has now bred a variety of rice that has two heads of grain whereas previous varieties had just one. The claim is that such varieties will feed the growing masses and stave off starvation. Yet again, nutritional quality is likely to be sacrificed for bulk production, and were such double-yielding varieties to sweep the world, the likelihood is that, even were bellies filled, a greater proportion of the world's swelling population would suffer malnutrition, thus perpetuating what we have already encountered with today's green revolution.

The market economy, combined with growing consumer aspirations worldwide, is beginning to have a major impact on natural resources and global systems, such as climate, especially through emissions of greenhouse gases. And as economies grow, particularly throughout the developing world, the resources used on average by each person and the way those resources are exploited become more of an issue than just increased population. In that respect, the United States is by far the most profligate nation in the world. With one twentieth of the world population, the United States consumes in the region of three quarters of the annual production of the world's natural resources.

Energy Use

Since the mid-nineteenth century, energy use worldwide has increased thirtyfold and today the average rate is about twelve million million watts (12 terawatts). The population then was approximately one billion, twenty per cent less than today. Each person's energy consumption on average has therefore increased sixfold over the past century. Yet vast differences exist between average consumption in the industrialized developed world and the developing world. The western European average energy consumption is nearly one half that of the United States. If the world had average European consumption, then total energy demands would be three times greater than today.

Energy intensity has reduced by more than half since the 1880s, and developing countries are expected to adopt the more efficient energy technologies apparent in industrialized countries in the 1990s. However, while improved energy efficiency may help to reduce the growth in greenhouse gas emissions, the very *development* of the developing countries will inevitably lead to huge increases in the total energy generated and consumed worldwide.

Agriculture and Climate

Forest decimation
Until the beginning of this century, the tropical forests of the world remained largely intact. They have since been decimated, for timber, agriculture and cattle ranching. During the 1980s alone, we destroyed eight per cent of the remaining tropical forest. Worldwide, tropical forests are now disappearing at an annual rate of 15.4 million hectares; an area twice the size of Austria. Should we continue, all of these forests will have gone in little more than a century. 1997–98 will probably be deemed the worst year ever for the wholesale destruction of forests, owing to fires started by humans running out of control because of lack of rain.

Mountain forests have fared no better than ones in tropical lowland, with vast areas completely cleared, leaving behind highly eroded, soil-less surfaces that can barely support any vegetation at all. Springs, streams and rivers, that once provided plentiful fresh, clear water, either no longer flow, or if they do, turn to torrents and flash floods in the rainy season, carrying away millions of tonnes of soil.

In the Tropics, gully and sheet erosion are rife once the trees have gone. Meanwhile, the soil, carried into the ocean, smothers mangroves and corals, and coastlines are put at risk from storms and sea surges. A succession of photographs of hilly regions that were once forested, and then cleared, shows how, over a few decades at most, a place of luxuriant vegetation can turn into a desertified ruin. Madagascar, Oaxaca in Mexico, parts of the Indian Himalayas, and the Andes in South America, all show signs of reckless land abuse.

Tropical forests contain up to ninety per cent of the world's species, and, according to Harvard biologist, E.O. Wilson, as many as 140 species may be lost each day because of deforestation. One estimate for Colombia puts the loss at two species an hour. Such forests also act as an important carbon store. According to Nobel

Laureate Paul Crutzen, of Germany's Max Planck Institute, the burning of biomass in the Tropics, including the annual burning of bush and savannah grasses, releases between 1.8 and 4.7 billion tonnes of carbon into the atmosphere each year, which is thirty per cent or more of total carbon emissions from human activities.

The growing world market economy, if not directly, is largely responsible for much of the current destruction of forests and soils. Tropical forests meet nearly 33 per cent of world demand for logs, twelve per cent of sawnwoods, and sixty per cent of plywood and veneer, with South-East Asia supplying about half the tropical timbers.

Ghost Acreages

In the 1970s the Swedish agronomist, Georg Borgstrom, exposed the fallacy of ascribing increased farm yields to better husbandry, pointing out that such yields were dependent on what he called non-visible or *ghost acreages* that existed in other parts of the world — usually in developing countries. An update on ghost acreages suggests that people in upper income countries require 4–6 hectares of land in continuous production to maintain their lifestyles. If the current world population of close to six billion had such standards the total land needed would be 26.5 billion hectares, which is double the Earth's land surface. Only 8.8 billion hectares of the 13 billion total exists as productive cropland, pasture or forest.

Meanwhile, the opening up of new lands for farming has led to massive increases in the amount of wind-borne dust. Joseph Prospero and his colleagues from the University of Miami estimate that up to 3 million tonnes of mineral dust are blown into the atmosphere from deserts and soils each year. Inez Fung and her colleagues at the NASA Goddard Institute for Space Studies in New York, estimate that soil dust from cultivation and deforestation may reduce surface warming by as much as one watt per square metre, just under half the warming caused by anthropogenic greenhouse gases.

Impact of Global Warming

Models used to justify a business-as-usual approach are fundamentally flawed because they treat the Earth's land surface as it would be had we not destroyed great tracts of natural vegetation. The models therefore ignore the impact of global agriculture on climate, and fail to account for the possible synergies and positive feedbacks caused by the destruction of the natural world (see Chapters 12 and 13). In addition, the models underestimate the impact of agriculture on the release of greenhouse gases, such as nitrous oxide and methane, not least their effect on the ozone layer. Nor do they take into account the devastation that would be caused by long and far more frequent heatwaves and other extremes of weather.

Drought, Heatwave and Flood

It is all very well to talk of average rainfall decreasing or increasing, as if we can adapt to and accommodate such trends. The decrease in rain may be associated with searing summer heat or, vice versa, the increase in rainfall the result of cloudbursts that wash away soil and crops in specific areas, at specific times. Lack of rain and scorching Sun, for instance, caused the 1988 drought in the southern states of the US, which cost the federal government over $3 billion in direct relief payments to farmers. Equally the deluges during the spring and summer of 1993 led to unprecedented flooding of the Mississippi and Missouri rivers, with several billion dollars worth of damage to farm produce and property.

According to the US Geophysical Fluid Dynamics Laboratory, warmer, drier summers over the US Great Plains, Western Europe, Northern Canada and Siberia — because of high carbon dioxide concentrations — would lead to soil drying out by at least twenty per cent during the crucial growing season.

In their chapter 'Heatwaves in a changing climate,' (see *Climate, Change and Risk)*, Megan Gawith, Thomas Downing and Theodore Karacostas look at the likelihood of increased heatwaves from global warming for two locations, Oxford and Thessaloniki. By the middle of the next century they anticipate — on the basis of the UK Met Office's climate models — that, over July and August, Oxford, and southern England as a whole, will experience at least ten times more heatwaves compared with the present. In Greece, not only is there likely to be a doubling of the length of heatwaves, from 50 to 100 hours, but whereas 50 hour events are relatively rare now, 100 hour events will become the pattern for most years. Such heatwaves, combined with drought conditions, will clearly play havoc with most crops.

1976 is synonymous with drought in the UK. In 1991, the UK Department of the Environment, as it was then, estimated — based on business-as-usual greenhouse gas emissions — that the probability of a summer as hot as 1976 would increase a hundredfold, from 0.1 per cent to 10 per cent by 2030, and to 33 per cent by 2050, which is once every three years. Such a high incidence would clearly change the face of farming.

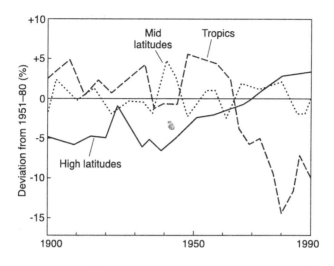

Figure 22. Rainfall trends in the northern hemisphere. (Source: New Scientist, *March 1995).*

Based on the UK Met Office's high resolution general circulation model, Paul Brignall and his colleagues at the Environmental Change Unit of Oxford University have investigated the likelihood of severe agricultural drought affecting Europe over the next fifty years because of global warming. The model indicates that by 2050, all Europe south of 48° N will become drier, and above that latitude most regions will receive more rain.

Drought leads to water stress in plants. Heatwaves, above 40°C for instance, lead to wilting and death, because of structural damage to essential proteins. According to Fitter and Hay temperatures of above 45°C, lasting for half an hour or more, will damage most plants. High soil temperatures will also damage roots and prevent nutrient uptake. If hot, dry conditions prevail during a sensitive time in a plant's maturation then the result could be catastrophic for yields, even though the adult plant survives.

Already global warming is having an impact on agriculture. Since record keeping began in 1866, 13 of the warmest years have all occurred during the past 20 years. The 1988 drought caused the grain harvest per hectare to fall by 22 per cent from the year before, with the result that for the first time in recent history US grain production dropped below consumption. The United States is responsible for nearly half the world's grain exports. Such losses are therefore of major concern and if the US's yields, particularly of corn, flounder because of heatwaves and disturbed patterns of rainfall, the world will be in trouble, especially in Asia, where imports are soaring because of the loss of land to urbanization, and the ever-greater demand for meat in the diet.

Sea Level Rise and Storm Surges

Sea level rise is not something that manifests itself all of a sudden and takes us by surprise. Nevertheless, its slow, inexorable rise will have a major impact on agriculture by the inundation of low-lying coastal regions, and salt intrusion of coastal aquifers. Owing to a number of powerful feedbacks, sea level rise is likely to be considerably greater over the next century than calculated by the main climate models, such as that used by the Met Office. A rise of one metre — all possible because of the thermal expansion of sea water without any glacial melting (see Chapters 3 and 6) — would threaten one third of the world's current cropland.

Rainfall and Water Stress

On average, each of us needs about one million litres of freshwater a year for all our requirements, including food production, which takes up about three-quarters of the whole. One third of the world's population are now living in countries which suffer from periodic shortages of water, with serious consequences for agriculture. Without taking climate change into account, the anticipated increase in population by 2025 could lead to two thirds of the world's population living in countries that suffer water stress. Any country which uses twenty per cent of its available water supply suffers from water stress, and from severe water stress when the ratio of water used goes up to forty per cent. By applying these criteria to the latest Hadley Centre model, Nigel Arnell, of Southampton University, predicts that by 2050 global warming will cause an additional 66 million people to be living in water stressed countries, with an additional 170 million people living in countries that are severely water stressed.

What about the impact on crops? As David Pimental notes, water is the primary limiting factor for crop production worldwide. To produce eight tonnes per hectare requires access to five million litres of water — the equivalent of 500 millimetres of rain. Wheat and other grains require one thousand litres to produce one kilogram, which is multiplied by one hundred to produce a kilogram of beef. Even so the amount required is only a fraction of that needed to fall as rain. For maize the amount needed as rain is approximately ten million litres per hectare, double the plant's basic requirements. But even that would be insufficient if surface temperatures rise and the rate of evaporation from the soil surface increases.

Irrigation: reaching its Limits

Although used on seventeen per cent of the world's total cropland irrigation enables agriculture to supply forty per cent of the world's food. With two or even three crops in a year now feasible through providing water on tap, irrigation has become crucial for food security. After a rapid expansion during the 1960s and 1970s, the area dedicated to irrigation reached 250 million hectares by 1994, but the rate of growth has now slowed and, at best, is un- likely to exceed more than 0.3 per cent a year over the next half century.

In many instances, aquifers are becoming depleted or salinized as salts accumulate; moreover, as rapidly as new areas for irrigation are opened up old areas are being discarded. Costs too are increasing rapidly as water tables fall. According to Pimental, to pump water from a depth of no more than 30 metres, needs three times more fos- sil fuel energy for corn production than for rain-fed production of the same amount. Meanwhile, over 25 per cent of the land used for irrigated agriculture in the Indus Basin is now damaged because of salinization and waterlogging.

An obvious consequence of global warming will be to increase demands for irrigation water without necessarily an increase in yields. A study on irrigation requirements in the United States indi- cated that even if precipitation increased by ten per cent, an in- creased surface temperature of 3°C would raise water requirements. Temperature increases with a fall in precipitation would therefore have a considerable impact on the availability of water.

Dry river basins, where the river and tributaries flow through semi-arid and arid regions, such as those of the Nile and Zambesi, will be particularly affected by climate change. Wet river basins, in the humid tropics less so. Changes in the pattern of rainfall rather than in total precipitation, can lead to major episodes of flooding and crop destruction, especially when the river systems have little extra capacity for storage. Heavier downpours, anticipated with global warming, can be extremely destructive, causing landslides and soil erosion.

Soil Erosion

Worldwide 75 billion tonnes of soil are eroded each year, with at least sixty per cent of that washed away into rivers or out to sea. According to Pimental, that loss in nutrient terms is equivalent to several billion tonnes of fertilizer — nearly the total amount of fertilizer applied each year across the globe. In the United States alone the fertilizer loss in terms of nutrients equals $20 billion. Higher soil temperatures will increase the rates of oxidation and thus the loss of nutrients and organic matter. Less organic matter means less soil organisms such as earthworms and insects which do so much to improve the ground. The soil is likely to compact more and the rate of run-off will increase, adding to potential problems of water stress.

Global Warming and Pests

Modern agriculture offers a rich harvest for the tens of thousands of pest species, as well as weeds, which wait for ground to be cleared. According to Pimental, despite the annual application of 2.5 million metric tonnes of pesticides, more than forty per cent of world food, forage and fibre production is currently destroyed by a combination of pests, plant pathogens and weeds. The loss is estimated at $500 billion per year.

Within the context of modern industrialized farming, global warming and warmer temperatures, with milder winters in temperate zones, will lead to a surge in pathogens and pests. Not only will some pests be able to take advantage of rising temperatures to spread to higher latitudes and altitudes, but also to increase their rate of reproduction by adding an extra generation. During the growing season some insect pests can produce 500 progeny per female every two weeks, and as many as 3,000 in a single generation. The European corn borer is able to produce four generations a year. A 1°C rise in temperature will enable the corn borer to extend its range northwards by as much as 500 kilometres. Locust swarms may become common in southern Europe. A 3°C rise would see a major expansion in insects such as the tobacco cut-worm, southern green stink bug, rice stink bug, lima bean pod borer, soyabean stem gall, rice

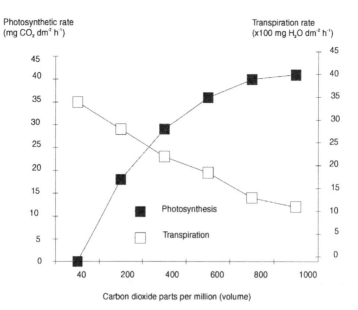

Figure 23. The graph shows the results of experiments on the effects of varying concentrations of carbon dioxide on the growth of the American poplar, as measured by photosynthetic uptake of the gas. The growth rate is found to increase with carbon dioxide concentration up to approximately double today's atmospheric concentrations. The efficiency of water use also improves with increased carbon dioxide levels.

weevil and soyabean pod borer. Animal diseases such as African swine fever are also likely to jump countries in a warmer world and may break out as far afield as North America.

In general, losses to insects and mites are higher in warmer regions of the world. As Pimental points out, potato growers in Maine average losses of around six per cent to insects, whereas in Virginia the losses rise to fifteen per cent, despite the use of more insecticides. As the world warms, the losses to insects could double to thirty per cent of production. Fungal and bacteria pests in plants will also thrive in a warmer world, especially if it is wetter. Mild winters, which are now becoming the norm in Britain, encourage outbreaks of fungal diseases, such as powdery mildew and strip rust in cereals, as well as potato blight.

Crop Response to Global Warming

Agronomists are quick to point out that the impact of global warming on agriculture will be tempered by the response of vegetation to increased levels of carbon dioxide, which may cause plants to photosynthesize more efficiently and as a result to grow more vigorously. This effect is especially true of C3 plants, which include temperate grasses and cereals such as wheat and rice, but is less true of C4 plants that include maize, sorghum and sugar cane. Of the 86 plants that comprise ninety per cent of per capita food supplies worldwide, 80 of them are C3 plants.

C3 plants are so-called because their first product of photosynthesis is an organic compound with just three carbon atoms. When C3 plants are exposed to the carbon dioxide levels now found in the atmosphere — approximately 360 ppmv — they burn off some of the carbon that has been reduced during photosynthesis in order to make more carbon dioxide available inside the leaves. A proportion of the photosynthetic gains are therefore lost to the plant. However, as carbon dioxide concentrations in the atmosphere increase, less burn-off is required to obtain the requisite concentrations of carbon dioxide, and stops completely once atmospheric concentrations have reached 1,200 ppmv — about four times pre-industrial CO_2 concentrations.

This improvement in efficiency of C3 plants as atmospheric CO_2 levels rise translates into increases in growth which, if channelled into seeds, should mean higher yields. C4 plants, on the other hand, operate more efficiently at today's relatively low carbon dioxide levels, and have less to gain in terms of rising atmospheric levels of the gas.

In addition, many C3 plants are weeds and a likely consequence of increased carbon dioxide levels will be to stimulate an epidemic of aggressively growing weeds. That would threaten the yields of C4 crops such as maize, sorghum, millet and sugar cane and lead to raised costs in weed control.

When carbon dioxide levels increase in the atmosphere the stomata tend to shut down. Water, which is normally transpired through the stomata is therefore conserved. Hence, increased carbon dioxide has a twin effect, especially in C3 plants, first to increase net primary productivity and second to conserve water. However, transpiration also serves another important function in keeping the surface of the leaf cool. Therefore, when temperatures rise in a world of higher atmospheric concentrations of carbon dioxide, the leaf surface could overheat, causing significant reductions in plant growth.

In their modelling of the potential impact of climate change on agriculture, Cynthia Rosenzweig, of the Goddard Institute of Space Studies in New York, and Martin Parry, of the Environmental Change Unit at Oxford University, assume that an increase in carbon dioxide, from 330 to 555 ppmv, will increase net photosynthesis in C3 plants by as much as twenty per cent, with associated gains in water use of around fifty per cent. However, any such gains must be offset against losses in the overall ecosystem, which includes the enhanced activity of soil organisms that are benefiting from the increases in the total amount of organic carbon generated.

The Efficiency of Carbon Dioxide Inspired Growth

In their paper, 'Dynamic responses of terrestrial ecosystem carbon cycling to global climate change,' in *Nature* (Vol 393, 21 May, 1998) Mingkui Cao and F. Ian Woodward have modelled the physiological implications of climate change on the terrestrial carbon cycle. Using the UK Met Office's general circulation model to predict climate change over the coming century, with an assumption that carbon dioxide will increase at a rate of one per cent per annum, they then look for broad consequences for vegetation and how much carbon is likely to be sequestered from the atmosphere.

With global warming and the physiological impact of increased carbon dioxide operating together, the model shows that between 1860 and 2070, as much as 309 billion tonnes of carbon should have accumulated in soils and vegetation across the planet. If that were to happen, the amount of carbon sequestered by net gains in biomass growth would equal fifty years of emissions at the rate of 6 billion tonnes of carbon a year. The model actually indicates that 58 billion tonnes of carbon should have accumulated during the period 1861-1990, therefore accounting historically for about two thirds of the carbon from anthropogenic sources that has found its way into terrestrial biomass.

The model tells us that we should not be over-worried by a doubling of atmospheric carbon dioxide and its impact on surface temperatures. Even so, once carbon dioxide exceeds 600 ppmv and global temperatures continue to rise, the stimulus for higher growth actually vanishes and net gains turn to net losses in the accumulation of carbon. Furthermore, such computer experiments take no account of anthropogenic changes to the landscape, including the transformation of land for agriculture through the destruction of forests. The reality is that the conversion of lands to agriculture leads to massive losses of soil carbon to the atmosphere, thus enhancing anthropogenic emissions.

Today, one third of the terrestrial surface, including the steep slopes of mountains, has been converted for human use, mostly through the degradation and destruction of forests, and many of these changes have taken place in the past forty years. Furthermore, the increase in greenhouse gases, especially carbon dioxide, is likely to be four times pre-industrial levels rather than double as used in their model. Consequently, the average surface rise in temperature will be far greater than the 4°C maximum predicted by the IPCC. Such changes will inevitably lead to severe disruptions of normal climate systems, with a tenfold likelihood of devastating heatwaves and storms. It is nonsensical to imagine that vegetation will draw down more carbon from the atmosphere than will be emitted from soils and decaying forests.

The Problems of Present Day Agriculture

Even without climate change, all is not well with agriculture today. The great post World War II surge in food production is grinding to a halt and yet the world's population will increase by approximately one billion people in just over a decade. According to the World Health Organization, as many as 3 billion people are currently malnourished. Yet, as David Pimental points out, worldwide per capita gains have been declining since 1983, not only because of population growth, but more significantly because of loss of land to degradation.

According to the Food and Agriculture Organization (FAO) cereal production will continue to grow by about one per cent per year — a doubling over the next 70 years — with increases in global production from 1,800 million metric tonnes in 1990, to 3,500 million tonnes in 2050. The combination of that, plus global trade and the supposed enhanced purchasing power of the world's population, because of industrialized development, is the recipe, according to such economic models, for better access to food and the banishing of hunger. There is no mention of the impact of global warming in this projection.

When modellers include the impact of climate change, 36 million more people are seen to be at risk from hunger in 2020 compared with the FAO figures. That number accelerates rapidly as global warming takes hold in the mid to latter part of the 21st century and, by 2060, could be as many as 350 million more than anticipated by the FAO. According to the UK Met Office's Hadley Centre model by 2080, as many as 40 million more Africans could be at risk from climate-induced food shortages. These estimates depend on population increases as forecast by the United Nations and on the impact of global warming as predicted by the models. As pointed out, the models understate the problem.

Moreover, through the need to average climate data over a 30-year period in order to minimize the ripples caused by freak events we will have introduced a delay in the calculation, just when changes are beginning to take off. In this respect how seriously can we take pundits such as Cynthia Rosenzweig, and Martin Parry? In their 1994 paper in *Nature* they tell us that farmers should be able to adapt in time to global warming, so much that overall yields will barely decline from the optimistic forecasts of the FAO.

The adaptations they propose concern industrialized agriculture and include changes in the date of planting, to profit from milder soil temperatures in high latitude zones; changes in the varieties and crops used; and finally changes to the irrigation and fertilizer regime. They assume in their model that irrigation water will make up for any major shortfall in water available to crops, and that irrigation will be one hundred per cent efficient, irrespective of the impact of climate change. When we look at the state of irrigation today, with more land coming out of production than going in; with rapidly falling water tables because of over-exploitation; and the extent to which crop yields are being slashed because of waterlogging and soil salinization, we can only conclude how dangerously optimistic such projections are.

In a 1990 study, the United Nations estimated that agricultural mismanagement had damaged more than 38 per cent of today's cultivated area — 552 million hectares — since World War II, and that overall, between 5 and 10 million hectares a year were being lost. One hundred years at that rate would leave the world with a fraction of the land for agriculture that it has today.

We are also losing land through the expansion of cities and urban sprawl. In 1994, Gary Gardner of the Worldwatch Institute, reports that Indonesia lost 20,000 hectares of cropland to urban expansion

— land that could have fed 330,000 of its citizens. In the United States cropland equal to two New York Cities — 168,000 hectares — were paved over each year between 1982 and 1992, while, in the same period, California lost more than three per cent of its cropland to urban and non-farm use.

In 1981, for the first time in the post World War II era, the amount of land in production worldwide began to decline. In terms of population, the amount of land harvested for grains has come down steadily from approximately 0.23 hectare per person in 1950, to 0.12 hectare in 1996. Taking those projections further into the future, the grain area per person will drop to 0.10 hectare by 2010, and to 0.09 by 2020, declining to 0.05 hectare per person within a century.

According to Lester Brown of the Worldwatch Institute, we are now close to the point when two trends — one, the limits to worldwide growth in the expansion of the food base, and the other, the rising expectations of the world's population to eat meat products — will cause real food shortages. Between 1990 and 1996, China's grain consumption leapt by 40 million tonnes with most being used as animal feed — a fivefold increase from 1976. Currently one third of the world grain harvest goes to the feeding of livestock rather than to the direct feeding of people.

Cities and Local Climate

Cities generate their own distinct climate that sets them apart from the surrounding countryside. Haze, smog, dust, particulate matter; all these have profound effects on the interchange of heat and moisture between the city and the surrounding atmosphere — tall buildings create swirling vortices; asphalt and concrete replace green vegetation, affecting the amount of light absorbed and reflected; impermeable surfaces send rain into storm gutters and subterranean culverts, leaving dry, parched surfaces.

The needs of cities — for freshwater, power, and provisions — have ramifications that go far beyond city boundaries. Cities need large volumes of water to flush out sewage and other wastes; they need an ever-growing network of roads to handle the flow of people and goods; they need massive amounts of energy to power the city; and they need increasing areas of farmland and forests to supply food, clothing, construction material as well as pulp and paper for the ever-expanding market in reading material. When all these factors are taken into account, a city the size of London requires an area of 19 million hectares — three-quarters the total land area of Great Britain. On its own Britain could not support its cities; it has to import energy in the form of fossil fuels and uranium, food, timber, cotton, pulp and paper, from all over the world. Even production on Britain's farms is tied in to the import of grains, fishmeal and oil seeds.

Metabolism of the City

Cities have a metabolism of their own, generating and using energy in vast quantities. As a result, cities tend to be warmer than their surroundings and to form their own microclimates, with city temperatures sometimes as much as 5°C higher than the surrounding area. More than 155,400 km² of the United States is now paved; an area equivalent to ten per cent of all arable land in the US which could produce enough grain to feed 200 million people. Instead, the world's motor vehicles consume half of the world's current production of oil, and apart from toxic pollutants, generate nearly twenty per cent of greenhouse gas emissions.

Buildings in cities, especially high rise ones, tend to slow and deflect winds. Studies of wind patterns and rainfall in St Louis, which is located in an area of flat farmland in the Ohio River Valley, indicate that the city is like a stone in a river, which upstream holds back the flow and downstream accelerates flow. Meanwhile, the abundance of concrete and light-coloured buildings, reflects sunlight, thus having a cooling effect compared to pastures and forests. Nevertheless, vegetation intercepts sunlight and encourages evapotranspiration, and when the affects of urban albedo are compared with latent heat cooling, the city tends to warm more quickly than surrounding countryside. The difference in the warming of the city heat island and the countryside generates a sea breeze circulation with the warm air of the city rising and drawing cooler, moister air from outside. Cumulus clouds tend to form which can result in heavy downpours. Between 1951 and 1960, for instance, London had 20–25 cm more summer rain than rural south-east England. That contrast appears to apply to most large conurbations compared with their surroundings.

If the heat island effect of cities is not taken into account, it can lead to a bias in recording the increase in temperature caused by the greenhouse gases. The growth of cities in the United States over the past century may have pushed up average temperatures by as much as 0.4°C.

Los Angeles: an Example of Inefficiency

One hundred years ago, Los Angeles was a small, rural town, set amidst rich farmland. It now covers 1200 km², with an intricate network of roads and superhighways. The city boasts the greatest number of vehicles per person in the world, with many of the population driving up to 130 km a day to work. Each day, the cars consume 30 million litres of petrol and generate 12,000 tonnes of pollutants. Worldwide, motor vehicles are now responsible for at least twenty per cent of all human emissions of carbon.

Traffic and Air Pollution

Despite attempts to control exhaust emissions, Los Angeles is smothered in smog. The Californian standard for ozone was exceeded at all monitoring sites in 1990 and in some areas the standard was exceeded in one day out of three throughout the entire year.

Motor vehicles jamming city centres have become one of the main pollution problems of the late twentieth century. With the rare exception of Curituba in Brazil, where a public transport system caters relatively efficiently with the needs of two million people, most large cities in developing countries, especially megacities like Bangkok, Manila and Sao Paolo, have tried to put in place *ad hoc* policies that still allow the private vehicle access to the city centre. The result is traffic bedlam, and in Bangkok, with as many as 500 new cars entering the city's road system every day, the average vehicle speed during the rush hour is a little over one kilometre per hour.

Rather than ban cars from city centres, most city authorities try a combination of measures to limit air pollution, such as improving public transport by adding to the network and providing special routes such as bus lanes. In addition, vehicles may be subject to tests as part of licensing, to improve safety and reduce emissions. Catalytic converters help reduce nitrogen oxide and carbon monoxide emissions as well as lower discharges of hydrocarbons. The quality of fuel makes a considerable difference; unleaded petrol with a low sulphur content is increasingly used to reduce lead and sulphur dioxide levels in ambient air.

Relatively simple measures can make a fundamental difference to air quality. Until the 1950s many British cities lost up to half the incoming solar radiation during the winter, as a result of people burning coal. Particulate matter, which acted as cloud condensation nuclei, exacerbated the effect of smog. Compared to the country-side, the city centre could lose as much as 45 minutes of sunlight each day throughout the year. In December 1952, a temperature in-version settled over London for five days, which held down a thick, chemical-rich, fog. 12,000 more people died from breathing diffi-culties during the next months compared to the same period in pre-vious years. The conclusion, that the fog had killed people,' was followed by Parliament passing the 1956 Clean Air Act, which banned the burning of coal in the grate. If people insisted on open fires, they had to use smokeless fuel, which had the sulphur taken out of it, as well as generating less smoke when burnt. Meanwhile, the electricity board established cheap nightly rates for central heat-ing purposes and therefore provided alternatives to open fires. Within four years of the 1956 Act, smoke emissions over London fell by more than 50,000 tonnes.

Health Effects of Pollution

A change in the chemistry of fuel can make a significant difference to emissions, that is if the vehicles are modern and fitted with de-vices such as catalytic converters. Nevertheless, Suspended particles less than ten microns (micron = one millionth of a metre) in diame-ter, (PM10) remain a serious problem and are now deemed respon-sible for major respiratory problems leading to death. In its third report on airborne particulate matter, the UK's Quality of Urban Air Review Group stated that 86 per cent of London's particles came from road transport. To meet air quality targets of 50 mg/m^3, these particle emissions need to be cut by two thirds. In London the limit was exceeded on 139 days between 1992 and 1994, and by 46 days in 1995, as well as to a lesser extent in other urban areas. Diesel cars are particularly to blame.

Nitrogen dioxide is four times more toxic than nitric oxide. Yet the oxide, while also being implicated in the production of ozone and other photochemical products, readily oxidizes in bright sun-light to the dioxide. In the lungs nitrogen dioxide affects the activi-ties of lymphocytes and other white corpuscles, as well as promoting inflammation and even pulmonary oedema, especially in those with chronic bronchitis. Sulphate aerosols are also damaging to the respi-ratory passages and lungs, especially aggravating asthma. Benzene, formaldehyde and polycyclic aromatic hydrocarbons (PAH), such as

benz-a-pyrene, are either released during the combustion of petrol or
are formed in the atmosphere as a result of photochemical reactions.
All are extremely toxic at low concentrations and aromatic hydrocarbons have been implicated in cancer of the bladder.

Ozone can cause inflammation of the lungs and a lowering of resistance to pulmonary infections, as well as acute changes in lung
physiology with possible effects on other organs such as the liver.

Suspended particles have begun to assume increasing importance
in terms of the damage they do to health. The particles consist of a
conglomerate of different substances that may be carcinogenic. Various studies suggest that for every ten micrograms per cubic metre
above the norm, mortality among an exposed population increases
by three per cent. A World Health Organization committee, headed
by Robert Maynard, head of air pollution in the UK Department of
Health, maintains that no level is safe. The committee also asserts
that three days with particle concentrations at fifty micrograms per
cubic metre will cause considerable extra suffering. Such exposure
would cause 1400 extra people to require their asthma inhalers, six
more than usual to need hospital treatment and four more to die, out
of one million. Every increment in PM10s of fifty micrograms per
cubic metre would lead to a doubling of those numbers, with twenty
four extra people dying when the concentration reached two hundred
micrograms per cubic metre. The transport industry finds such numbers contentious, yet during a bad smog episode in London between
12 and 16 December 1991, an extra 160 people died.

The lesson of Los Angeles with regard to car use and pollution
does not seem to have been readily learnt elsewhere in the world. On
the contrary, most governments bend over backwards to accommodate private cars in their city centres.

According to the department of the Environment (now the Environment Agency), between 1984 and 1996 the number of cars in the
UK increased from just over 16 million to 20.5 million. In 1995,
road traffic in the UK was responsible for 75 per cent of total emissions of carbon monoxide (London 99%), 46 per cent of nitrogen
oxides (London 76%), 29 per cent of hydrocarbons (VOC = Volatile
Organic Carbons, London 97%), and PM10s 26 per cent. Add air
traffic to motor vehicle pollution and the total effects are considerable, including a significant loss of sunlight.

Chapter Twelve: The Politics of Climate Change

THE evidence that human activities are causing climate change is now widely accepted. The United Nations, supported by governments, established the Intergovernmental Panel on Climate Change — the IPCC — in 1988. Since then, the IPCC has had a number of meetings to assess the extent to which humans have caused climate to change and, on the assumption that global warming activities will continue, are likely to bring still more change in the future. It has reviewed scientists' work from all over the world, including climatologists, meteorologists, atmospheric chemists and oceanographers. The IPCC has also called on governments to encourage climate research in an attempt to obtain consistency between historical evidence and the degree to which models can accurately predict present conditions from past data. The scientists agree that global warming, resulting from greenhouse gas emissions, is not just a theoretical concept, but is actually happening in front of our eyes.

Since the IPCC's first scientific report in 1990, governments have held a series of meetings, not least the 1992 Rio de Janeiro Conference on Environment and Development (UNCED) at which more than 160 countries signed the Framework Convention on Climate Change. Official confirmation of global warming came in 1995, when the IPCC published its Second Assessment Report, written and reviewed by some 2000 scientists. It stated that "the balance of evidence suggests there is a discernible human influence on global climate."

Scientific Projections

Rising temperatures are already the clearest sign of climate change. So far, according to the IPCC, global average temperatures have risen 0.6°C above the pre-industrial average. Nine of the hottest years on record have occurred since 1988; six of the first eight months of 1998 were the warmest since records began in 1866; and July 1998 was the hottest month ever. According to the IPCC's latest coupled ocean-at-

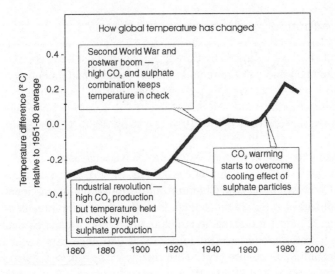

Figure 24. Temperature differences between 1860 and the present relative to the 1951–1980 average.

mosphere models, by 2080, for double pre-industrial carbon dioxide concentrations, we would see a global average increase of 2.5°C, with perhaps 4°C over land masses, particularly in the northern high latitudes, 3° to 4°C over parts of the Arctic or Antarctic, and possibly substantial regional variations from the global average. If the increases in temperature seem modest, it should be noted that a 3°C cooling brought on the most recent ice age. What is more, a second doubling of pre-industrial levels of carbon dioxide, which could occur by the end of the 21st Century, would lead to a catastrophic rise of up to 10°C.

The Third Conference of the Parties to the Framework Convention on Climate Change took place in Kyoto, Japan, at the beginning of December 1997. As a finale to the Conference, 149 nations adopted the Kyoto protocol that, when ratified at the Buenos Aires meeting of November 1998, would lead to an initial 5.2 per cent reduction in greenhouse gas emissions on 1990 figures, using carbon dioxide as the basis of measurement. Such reductions will have little impact on global warming; the IPCC's climate models show that, if adhered to, the cutbacks in emissions will lead to only a 0.1°C improvement on the 1.5°C increase expected over the next fifty years.

What Action?

The United States, followed by Europe and South-East Asia, are currently the main players with regard to fossil fuel use and greenhouse gas emissions. As the world's greatest consumer of fossil fuels per capita and in bulk use terms, the US agreed in Kyoto to cut emissions by seven per cent below 1990 levels by 2008. However, its emissions are currently thirteen per cent above 1990 levels and are set to rise to thirty per cent above over the next decade.

What hope is there of the US achieving its goals? David Sandalow, Chief Advisor to the Clinton Administration on international environmental issues, claims that climate change is 'a top presidential priority,' and enumerates a number of domestic initiatives that indicate Clinton's commitment to achieving the necessary reductions, including developing cars with the motor industry that have three times better fuel efficiency than current models. Clinton has also increased investment in mass transit systems and in renewable energy and energy efficiency programmes, bringing the total spent in 1999 to $1 billion. Houses are to be made doubly energy efficient and, aided by low cost loans, one million roofs should have solar energy panels by 2010. The total amount of electricity currently generated from renewable sources — wind, hydro and solar — is less than three per cent and the aim is to increase it to 5.5 per cent by 2010.

Critics of the Administration's record on global warming issues point out that such initiatives will make no more than a small dent in the US's rapidly rising greenhouse gas emissions. On the basis of business-as-usual the US continues to subsidize the fossil fuel industry by as much as $18 billion a year, while providing tax breaks for exploration, production and foreign royalties as well as military protection, at a cost of $57 billion a year, to ensure a continuous flow of oil from oil-rich states. In contrast to Europe, energy taxes in the US are not to be increased and fuel prices remain among the lowest in the world.

Article 17

At Kyoto, by pushing through Article 17, the United States got itself off the hook. Article 17 stipulates that countries may meet their obligations by trading emissions with countries that have legally binding greenhouse gas limits. Countries which are in *credit* because their emissions are sufficiently lower than their recorded 1990 levels can sell such credits to other nations, who can then offset their own excesses. The collapse of the Soviet Union and its economy has led to far less fossil fuel being burnt than in 1990. Under the Kyoto Treaty, by 2012 Russia is allowed to increase emissions by fifty per cent and the Ukraine by 120 per cent. In the meantime they can sell their respective credits to another country which can then do less to reduce its own emissions. By purchasing such credits and investing in forest projects that are considered to be valid greenhouse gas sinks, the US will be able to convert its notional seven per cent cut in emissions into a real increase of up to ten per cent. Nor are the countries that receive payment for their greenhouse gas credits obliged to invest in clean or renewable sources of energy. They are at liberty to spend on what they want.

Two other trading mechanisms from Kyoto involve joint implementation, whereby one country invests in another and can claim a carbon rebate for itself if the investment involves a greenhouse gas sink. The other is the *Clean Development Mechanism* whereby a country helps another to avoid burning more fossil fuels by offering its technology and know-how,

The industrialized nations are clearly responsible for much of the past burning of fossil fuels. Nevertheless, countries of the Third World are adding increasingly to the inventory of greenhouse gases, not least through land use changes, but also through their own programmes of industrialization. The Climate Convention, in fact, called on industrialized countries to cut back greenhouse gas emissions to 1990 levels by the year 2000. In the long run, that first step, if adhered to, will still fall short of stabilizing greenhouse gas build-up in the atmosphere. Stabilization of atmospheric concentrations of carbon dioxide at around today's level of 358 parts per million (by volume) would require emissions to be reduced worldwide by between 66 and 80 per cent of current levels.

Future Warming

An important aspect of the IPCC's analysis of global warming is to project a business-as-usual scenario as far as one hundred years into the future in order to predict might happen to climate if we continue emitting greenhouse gases without concern for their impact. The models used in such projections need to reflect accurately the interchanges of greenhouse gases between the atmosphere, oceans and soils, and their changing chemistry as they interact with one another under the influence of sunlight. Biochemistry, geochemistry and atmospheric chemistry all need to be modelled into the differential equations and then the physics of heat exchanges applied to circulation models of the atmosphere. Compared with the dynamic processes that underlie the real world, the models are manifestly simplistic, but they are the best we have.

Confidence is growing in the models and their projections following the reasonable fit between the observed data for global surface temperature between 1860 and 1990, and the results obtained from estimating the impact on temperature of the increase in greenhouse gases over the same period. The fit is reasonably close when the cooling effect of sulphate aerosols — some derived from volcanoes — is taken into account. The projections depend critically on emissions now and in the future as well as on the sinks, which take the greenhouse gases out of the atmosphere.

There is an implicit assumption in the GCMs that life will continue to function effectively as a vital, integral part of the carbon cycle, thus compensating against our profligate burning of fossil fuels. It is perhaps on this issue — of life carrying on as usual — that the IPCC's projections on the future impact of greenhouse gases on global warming are at their shakiest.

Past as well as present greenhouse gas emissions will affect future global warming, because of the time lapse for the gases to disappear from the atmosphere. Between forty and sixty per cent of the carbon dioxide currently released into the atmosphere, is expected to take about thirty years to be removed. The IPCC has taken different scenarios of future energy production and use into account, and whether they derive from fossil fuels, nuclear power or renewable sources.

Working Group Two of the IPCC is aware of the deficiencies in our knowledge about climate change. It has warned that our models are useless once carbon dioxide concentrations have doubled or more, with even more catastrophic consequences to our environment and hence to ourselves. The IPCC reminds us that:

> ecosystems contain the Earth's entire reservoir of genetic
> and species diversity and provide many goods and
> services critical to individuals and societies.

We rely on the natural world to provide us with food, fibre, medicines and energy, while taking for granted that nature is also involved in the processing and storing of carbon and other mineral nutrients, in assimilating wastes, as well as in regulating run-off and controlling floods, soil degradation and coastal erosion. According to Working Group Two:

> Unambiguous detection of climate-induced changes in most ecological and social systems will prove extremely difficult in the coming decades. This is because of the complexity of these systems, their many non-linear feedbacks, and their sensitivity to a large number of climatic and non-climatic factors, all of which are expected to change simultaneously. The development of a baseline projecting future conditions without climate change is crucial, for it is this baseline against which all projected impacts are measured. As future climate extends beyond the boundaries of all empirical knowledge... it becomes more likely that actual outcomes will include surprises and unanticipated rapid changes.

The scientific consensus is that global warming will bring about the greatest rise in temperatures in high latitudes, and the least close to the Equator. For the mid-latitudes an average global warming of between 1° and 3.5°C over the next century will lead to a shift of the temperature gradient (isotherm) by as much as 550 km polewards, or equivalent to an altitude shift of 550 m. Certain tree species are capable of migrating 200 km a century, and global warming will lead inevitably to dramatic changes in vegetation.

Energy Forecasts

The World Energy Council (WEC), with representation from the energy industry in over 90 countries, has the most up-to-date analysis of energy use. Taking into account the industrial and developmental aspirations of countries throughout the world, the WEC has come up with four energy scenarios (see Appendix I), the differences between them being the effort made by countries to curb future emissions of carbon dioxide through variations in the fuel mix and the level with which energy-saving measures are introduced.

The scenarios cover growth rates and fuel mix until the year 2020, and then extrapolate on for one hundred years until the year 2100. The base point year is 1990, when world energy demand amounted to 8,800 millions of tonnes of oil equivalent, or 12.1 terawatts, with the release of 6 billion tonnes of carbon from the burning of fossil fuels. The world population in 1990 was 5.3 billion. The United Nations has estimated population growing to 8.1 billion by the year 2020; then to 10 billion in the year 2050 and finally 12 billion by the year 2100, with ninety per cent of the growth in the developing world.

The WEC assumes that considerable efforts will be made in all four scenarios to improve energy efficiency, although to different degrees depending on the commitment to reduce greenhouse gas emissions.

Action against Climate Change

However inadequate, the climate models do indicate that our activities are probably responsible for recent warming and changes in climate. They also tell us that action is vital to offset any further warming. The main objective of the Climate Convention is for the stabilization of greenhouse gas concentrations in the atmosphere:

> at a level that would prevent dangerous anthropogenic
> interference with the climate system. Such a level should
> be achieved within a time frame sufficient to allow
> ecosystems to adapt naturally to climate change, to ensure
> that food production is not threatened and to enable
> economic development to proceed in a sustainable
> manner.

Such a statement raises questions about food production at a time of climatic uncertainty, the time required for natural ecosystems to adapt to climate change, and the nature of *sustainable development*. Is it achievable in the face of rapid population growth and human aspirations for consumer goods, including private motor vehicles? The scope for reducing consumption of natural resources is considerable, without having to alter current lifestyle patterns. By practising simple energy conservation and preventing extravagant losses, we can significantly reduce carbon emissions.

Flaring off gas at wellheads is one unnecessary emission of carbon gas. Enough methane escapes each year from leaks in Russia's gas pipeline to fuel a country the size of Britain. Euan Nisbet, a member of the European Commission's Methane Monitoring Unit, points out (*New Scientist*, 25 May, 1996) that the leaks amount to 6.5 per cent of total worldwide methane emissions. Plugging the leaks would therefore make a significant difference. However, the European Commission has cut funding for Nisbet's monitoring unit.

The Framework Climate Convention makes reference to the notion of the precautionary principle whereby measures should be taken: 'to anticipate, prevent or minimize the causes of climate change and mitigate its adverse effects.' The problem is how do we know for sure what the climate is likely to do and how effective our measures are likely to be?

The Convention tries to answer the problem by binding countries to a commitment to action, even though some uncertainty still prevails. It states:

> where there are threats of serious or irreversible damage ... lack of full scientific certainty should not be used as a reason for postponing such measures, taking into account that policies and measures to deal with climate change should be cost-effective so as to ensure global benefits at the lowest possible cost.

Cost-Benefit Analysis

In *Global Warming: the Complete Briefing,* John Houghton, who has been both Chairman and Deputy Chairman of the IPCC's Scientific Committee, makes various suggestions about how to evaluate any measures for reducing greenhouse gases.

He takes an analysis by William R. Cline who works at the Institute for International Economics in Washington, as a baseline for costs. Cline's starting point is the cost to the United States of coping with the global warming that may be set in motion over the next fifty years. The temperature rise is expected to be 2.5°C, and when all the costs are taken into account the amount comes to $62 billion a year — that is for: loss of land and dyke construction to combat sea level rise; loss of water supplies; losses to agriculture and forestry; increased sickness; the added cost of air-conditioning, and other contingency factors. That estimate is about one per cent of US gross domestic product (GDP).

Similar exercises in other developed countries have led to esti-

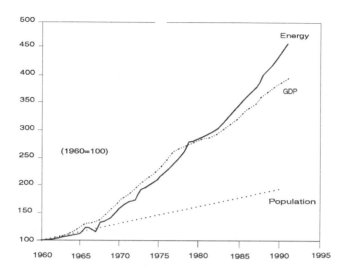

Figure 25. Growth in developing countries: Energy, Gross Domestic Product and Population, 1961–1991. (Source: British Petroleum; United Nations; World Bank; Population Reference Bureau).

mates of two per cent of GDP, and as much as six per cent of GDP in developing countries. The higher figure relates to the comparative poverty in developing countries, as well as their vulnerability to destructive events, such as cyclones that jeopardize the lives and livelihoods of millions of people, as in Bangladesh.

William Cline admits it would be wrong to put too much store by such GDP percentages of the cost of global warming. He warns that the analysis only covers the next fifty years, after which the effects of global warming could become more acute. An economic value cannot account for the loss of species, environments, amenities and indigenous lands and livelihoods.

In addition, a study by L.S Kalkstein, carried out for the United States Environment Protection Agency, indicates that a doubling in atmospheric carbon dioxide levels might lead to an extra 294 deaths a year per million of the US population. The number would reduce to 45 deaths per million if special measures were taken, such as building cooler, light-reflective buildings, and adapting behaviour to suit the new climate. The higher figure, applied to a world population of 10 billion, would lead speculatively to an extra 450,000 climatically-induced deaths per year.

Course of Action

In the light of the potential impact of global warming, economists are now engaged in a debate as to which is the best policy: one which advocates action now to prevent future change, or one which holds that it is cheaper to do nothing now, but pay up when necessary, in the hope that it will not prove too expensive.

The trouble is that the issues are muddled and future impacts can only be guessed at. In all probability tropical storms will worsen; crops will fail more often; diseases will take a greater toll than ever before on people and livestock; and life will generally get more difficult, especially in certain parts of the world. One country may be blessed with good luck and experience no damage from global warming, whereas another, like Honduras or Nicaragua, may be devastated. Is it better to take out insurance against damage, even though premiums are high? Or is it better to take concerted action, through agreements reached within the United Nations Framework Convention on Climate Change, to prevent damage where possible by building defence structures like dykes? Or is a mixture of both strategies a better way forward?

The choice will undoubtedly depend on an evaluation of the costs and benefits of different actions, or of a lack of action. At one extreme, economists claim that acting now to mitigate global warming will impede economic growth and be more costly than paying to repair damages later. Much better, they say, to develop a strong economy now so as to bear whatever costs arise later. At the other extreme, critics such as Aubrey Meyer and Anthony Cooper of the Global Commons Institute in London are saying that such economists have grossly underestimated the costs of global warming damage. They advocate action now and declare that the lion's share of the costs for climate change mitigation should lie with the developed, industrialized countries, since they are mainly responsible for unleashing global warming.

The Costs of Global Warming

Depending on the criteria used costs of global warming can vary dramatically. The IPCC's economists claimed that total global warming costs over the next century were unlikely to exceed two per cent of gross world product (GWP). However, the Global Commons Institute (GCI) found that costs are unlikely to be less than three per cent of GWP, and could be as high as 25 per cent.

GCI takes issue with IPCC's assumptions on economic impacts. Firstly, Working Group Three economists estimated global warming damages on the basis that the average temperature gain was 2.5°C rather than bracketing the damages to cover the full 1.5° to 4.5°C range given by IPCC's Scientific Committee. Secondly, they failed to include climate feedbacks that could make global warming more extreme. Finally, it takes issue with IPCC economists over their valuation of life in a developed country compared with a less developed one. They also challenge the notion that damage assessment of equivalent environments should be valued differently depending on whether they are in developed or less developed ones. Such distortions are clearly advantageous for the developed world should it agree to pay for damage limitation or reparation in other parts of the world. If, for example, a life in Bangladesh is evaluated at one fifteenth of a life in the developed world, then the industrialized countries would presumably be able to set aside less for possible compensation than they would for loss of life in their own countries.

And what about land? Is one hectare of wetland in the Florida Everglades, with its amenity value, worth exactly the same as a hectare of wetland in the middle of the Amazonian jungle? Who is to say that one is worth more or less than the other? Perhaps a way to judge would be to look at the respective ecological functions, for instance, how effectively each ecosystem fixes carbon through photosynthesis, or to evaluate ecosystems against a biodiversity standard, or by their ability to regenerate.

The purchasing power of a country's currency is also at stake. In their original estimate of costs, the IPCC economists calculated global warming damage on a basis in which local currency was corrected for purchasing power on the world money market and was divided by a gross national product that had not been corrected. This error resulted in an underestimate of the proportion of a country's income that would be spent on combatting the effects of global warming. Yet again, first appearances suggest that the cost of damages in less developed countries would be lower compared to equivalent damage in a developed country, such as the United States.

These factors all lead to a low evaluation of global warming costs. The distortions also indicate that one third of humanity living in the developed world will possibly be on the receiving end of twice as much damage as the rest of humanity living in the developing world.

Such a distortion is clearly nonsense, especially when we take into account that, for the most part, less developed countries are more vulnerable to global warming than the developed world. The corrections put in place by the Global Commons Institute go some way to redressing the balance. Not only are costs likely to be a far greater proportion of both GNP and GWP, but the proportion of damage likely to befall the developing world is twice that of the developed world. A sense of perspective is therefore restored.

We must therefore be cautious in taking IPCC's evaluation of costs at face value. However, on the assumption that the damages incurred by global warming currently lie between one and two per cent of gross world product, John Houghton divided that amount by the total quantity of carbon emitted into the atmosphere through human activities. One per cent of GWP amounts to approximately $200,000 million and the amount emitted is 7.5 billion tonnes per year. That gives a cost of $25 for each tonne of carbon emitted. If the cost of global warming is two per cent of GWP, then carbon emitted would cost $50, and $100 if it were four per cent of GWP.

Unit Carbon Costs

Houghton suggests that we should use such unit costs as a basis for evaluating the total cost of different levels of action. We might decide that we cannot afford to let carbon dioxide build up much beyond current levels, in which case we must cut back considerably on fossil fuel burning — perhaps by as much as two thirds. To prevent greenhouse gases exceeding an concentration of 400 ppmv by the year 2100, the annual cost would be at least three per cent of GWP. Setting levels for stabilization a little higher, at 450 ppmv, brings the cost down to one per cent of GWP, and at 500 ppmv to about 0.75 per cent of GWP.

However, we may not agree with the basic assumptions of costs, and instead believe that we should pitch them much higher. Then the costs of doing little or nothing now would rebound with far higher costs incurred later. The costs may escalate from three per cent of GWP when greenhouse gases levels are curbed to current levels, to as much as 25 per cent of GWP when gas levels are allowed to rise to 600 ppmv and above, in a business-as-usual scenario. Given our level of ignorance about the impact of raised levels of greenhouse gases on climate, we might be wiser to take the precautionary principle to heart and pay now rather than wait.

A particular problem with any assessment of the cost to society from an impact such as climate change is that the perpetrators — the industrialized countries for instance — are not necessarily the same as the victims. Future action therefore depends on international agreements to bring some notion of *equity* to bear on the argument, so that those who escape or benefit from the impact of climate change should help meet the costs of those who suffer.

What is Equitable?

The notion of equity is proving a thorny issue, because of the difficulty of pinning down the causes of environmental change. In today's world, all countries, whether *developed* or *developing* bear some responsibility for greenhouse gas emissions that stay in the atmosphere at the end of the year and add to the build-up. We have a good idea where the fossil fuels are burnt and in what quantities. However, the perpetrators of forest destruction may be poor peasants, and often the damage they do is equivalent in gross terms to the environmental damage perpetrated by those in high income countries. However, the peasant may feel that he has little choice if his family is to survive; his lands elsewhere may well have already been taken, for development projects, such as hydro-electric schemes, or industrialized agribusiness.

In the mid 1970s population biologist Paul Ehrlich, from the University of Stanford, and John Holdren, from Berkeley University, came up with the notion of total environmental impact of a country or population, as a product of population, its level of affluence and the damage wrought by the technology used to support its lifestyle. Affluence was to be measured simply by the *per capita* consumption of resources, and technology by the extent to which each unit of production incurred an environmental cost. Underlying environmental cost was the notion that more powerful technologies would be needed to provide more people with greater affluence — nuclear power stations for instance to generate large amounts of electricity.

How is environmental cost or technological efficiency measured? One suggestion is to use *per capita* energy use, to signify both environmental cost and the degree of affluence. In fact energy use can give only a crude guide to environmental impact or indeed to affluence. On average, people in the United States are twice as profligate in their use of energy as Western Europeans, yet they may be no more affluent, just more wasteful. An American, for instance, uses the same amount of energy as 3 Japanese, 6 Mexicans, 14 Chinese, 38 Indians and 530 Ethiopians. In addition, different energy forms have different impacts. A well-sited wind farm, for instance, may use less than half the land area for a given output compared with a coal-fired power station when coal mining is also taken into account. Consequently, we need to distinguish which form of energy is used and how. Affluence depends on the accessibility of energy. People feel reasonably *well-off* when they have electricity at the flick of a switch.

Fossil fuels currently meet more than 75 per cent of the world's energy requirements. Therefore, average *per capita* carbon emissions should give us some idea of a population's contribution to the build-up of greenhouse gases and to global warming. One approach in the pursuit of equity is to allocate equal *per capita* emissions; another to allocate emission amounts on a country by country basis so that penalties are incurred if the quantities are exceeded. Michael Grubb, at the Royal Institute of International Affairs in London, argued some years ago for an individual quota of one tonne of carbon per year. Since the yearly emissions of carbon from fossil fuels amount to six billion tonnes, Grubb's quota system, if applied now to the world population, would keep emissions close to current levels. An energy profligate nation such as the United States would therefore pay for its profligacy, or better still, be induced to curb its energy use and make units available elsewhere. By Grubb's system, the average American's carbon emissions are 5 times over their allocation, and the United States as a whole is consuming 1500 million carbon units above its allocation. Meanwhile, average Chinese emissions are half a unit, and China as a whole has 500 million units in hand. However, China's rapid economic growth and rising energy consumption in the form of coal, will soon mop up any remaining quota.

A potential failing of such a quota system is that countries with a high population density but low overall energy use, have an advantage over countries where relatively low population density but high energy use prevails. Perhaps, the quota system needs to be linked to a fixed population size.

Quotas

Another problem in maintaining equity could arise if a small country with relatively few resources of its own was on the receiving end of damages caused by global warming. Such countries would perhaps be better served through the distribution of aid from a central agency dedicated to the mitigation of climatic impacts. If nothing was done to mitigate global warming, then under a business-as-usual scenario, as many as 150 million people could be displaced by the year 2050 — about 3 million a year.

Two thirds of the displaced would be as a result of sea level rise, with the remaining third from drought affecting survival in regions where people still subsist. A cost of between one and five thousand dollars for resettling each of the three million displaced has been mooted. In fact, resettlement schemes for those forced to leave their lands, because of extensive hydroelectric projects, for instance, have an appalling record of unfulfilled promises. Consequently, global warming is likely to leave a trail of environmental refugees who will be abandoned by their own governments, and instead of receiving some form of compensation, will be written off as undesirable immigrants by other governments.

Achieving Equity

The Global Commons Institute has vigorously promoted the theme of equity in international negotiations on global warming and has come up with a radical approach of *contraction* and *convergence*. Aubrey Meyer and his colleagues accept that the only fair solution is that each individual currently alive has a right to energy use, and therefore to some inevitable carbon emissions. Given the aspirations of nations to achieve westernized lifestyles and patterns of consumption, equity will lead to greenhouse gas emissions far exceeding those of today and being more along the lines of the WEC's middle to high projections. *Contraction*, says Meyer, is for survival and means reducing global fossil fuel use by sixty per cent over the next one hundred years. That means that profligate users need to curb their emissions by efficiency gains, and by implementing alternative sources of energy. *Convergence*, on the other hand, will ensure that unequal *per capita* consumption of fossil carbon is *policy-driven* to parity.

For *contraction* and *convergence* to work, targets and schedules will need to be set for all countries, based on their current populations and emissions. A sixty per cent final reduction on 1990 levels would mean an entitlement of approximately 0.5 tonnes of carbon per individual. Forest destruction and its inherent carbon emissions would also need to be included in the calculations. Meyer views emissions trading as a useful interim mechanism since it will provide countries with an incentive to keep their emissions below the quota. Consequently a number of developing countries have shown interest in promoting the overall idea of equity. It is unlikely to be accepted by the United States, which would ultimately have to reduce carbon emissions by ninety per cent. Furthermore, the United States argues that it is only by its consumption of cheap fossil fuels that it can compete for trade in global markets against cheap labour.

Sustainable Development

In the years leading up to the Rio Conference in 1992, the Brundtland Commission espoused the notion of sustainable development in order to meet 'the needs of the present without compromising the ability of future generations to meet their own needs.' What, in fact, do we mean by *sustainable* and what do we mean by *development?* Underlying the advocacy of development is the idea that all of humanity should have access to the technological benefits of industrialization: hence to shelter, adequate food, fresh piped water, work, medicines, mobility and not least to consumer goods.

What has been the reality of post World War II development? Part of the problem in assessing development has been to find a valid indicator. Just as *per capita* energy use may indicate the average individual's impact on the environment in attaining his or her degree of affluence, so too *per capita* gross national product (GNP) has been used as a guide to the level of a country's economic and social well-being, the conclusion being that if people have reasonable incomes they can then afford the goods and services that make them comfortable. But, as the ecologist Robert Goodland, and the economists Herman Daly and John Kellenberg of the World Bank, point out, GNP is a fundamentally flawed measure of socioeconomic and environmental health. In deriving their accounts on national income, governments in general disregard the depreciation, degradation and even total loss of *natural* capital, which may include soils, forests, watershed areas, water quality, photosynthetic capacity and even minerals.

On the contrary, if jobs are generated in cleaning up operations, such as after an Exxon Valdez oil spill and more than $2.5 billion are spent, that all registers as a plus to national income. The wildlife lost as a result of an oil spill, or perhaps more seriously because of the destruction of habitat, is not accounted for; nor as yet is damage

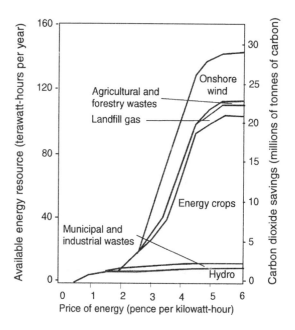

Figure 26. Renewable resources in the United Kingdom in 2005.

to the global environment, to the ozone layer for instance. In fact, the consumption of natural capital registers as income, rather than featuring as loss. Such practices are a complete misrepresentation of a country's economic well-being, and Goodland and his colleagues pillory it as 'the worst of all accounting sins.'

But that is not all. GNP accounting actually shows an economy is doing better when inefficient, resource-costly practices are pursued and, vice versa, efficient use of resources would appear to be *poor economics.* A person who walks or cycles to work therefore does less for the economy than another who uses public transport. A driver who sits alone in a large car in a traffic jam, burning fuel, contributes far more to national income than his or her more thrifty and perhaps health conscious fellow beings. Moreover, the ill health generated through traffic pollution adds even more to national income, on the assumption that medical care is available. A recent World Bank study of gross domestic investment in Mexico shows that as much as 25 per cent of GDP translates into a three per cent loss when account is taken of the depletion of natural resources, of man-made resources, in addition to environmental degradation.

GWP grew from \$3.8 trillion in 1950 to \$18.7 trillion in 1991. That more than fourfold increase in world product suggests that world affluence and well-being increased proportionately. The data tell a different story: in 1960 the wealthiest twenty per cent of the world population received incomes that were thirty times greater than the poorest twenty per cent: by 1990 the difference between the two groups had grown sixty times and the poor had become proportionately poorer.

The rationale, that we need to continue with conventional economic growth to improve the lot of humanity, does not seem to be working. Every year the gulf between the rich and poor is widening, while the environment becomes increasingly degraded. On average per capita incomes in industrialized countries are more than 20 times greater than those in low and middle income countries. If catching up is not feasible, then the only alternative in a world striving for equity is that those with extravagant lifestyles should substantially reduce their demands on natural resources and the environment.

Alternative Energies

Climate change is serving as a serious warning that business-as-usual may be endangering the future of humanity. But, what about using technologies that do less harm; for instance developing energy systems that emit minimal amounts of greenhouse gases? We can try and cut back on energy use by cutting out profligate practices; conserve energy in homes and businesses; turn to energy-saving appliances; develop renewable energy sources; vastly improve the efficiency of energy supply systems, like coal-fired power stations; develop biomass resources, as a renewable fuel resource; or press on with nuclear power. More probably, we can embark on a mixture of all these approaches. The question then is whether they will be sufficient to serve and enable growth and development to proceed?

Is Nuclear Power the Answer?

It might be tempting to replace a large emitter of greenhouse gases, such as a coal-fired power station with an equivalently sized nuclear generator. The argument is simple: a coal-fired power station with a generating capacity of 1GW of electricity emits 130,000 tonnes of sulphuric acid, and 32,000 tonnes of nitric acid, as well as 1 million tonnes of carbon in the form of carbon dioxide. A 1GW nuclear power plant operating over the course of a year emits relatively few of those gases.

However, nuclear power relies on fossil fuels for the extraction of uranium, often too for its enrichment into nuclear fuel, for construction, for safety and transportation of materials. Currently, in the United Kingdom, nuclear power operation leads to emissions of four per cent of the carbon dioxide than a comparable size coal-fired plant. In its pamphlet *Nuclear Power and the Greenhouse Effect,* the United Kingdom Atomic Energy Authority argues that the 1000 GW of nuclear power forecast to be operating in the world by the year 2020, would generate twenty per cent of the world's electricity and thereby save on the annual emission of more than 1 billion tonnes of carbon — approximately thirteen per cent of total carbon emissions today. Obviously an even more aggressive nuclear power programme worldwide would reduce greenhouse gas emissions still further while supporting industrialized growth.

But the advocacy for nuclear power has overlooked other factors, not least the massive capital investment required to safeguard nuclear wastes over many generations, as well as to prevent accidents on the scale of a Chernobyl. The investment required to construct and operate a nuclear plant safely has an *opportunity cost* in taking capital away from investment in other measures, such as energy-saving devices. In 1990 the energy analysts, Gregory Katz and William Keepin showed that investment in nuclear power in the United States was at least 7 times more costly than bringing about an equivalent saving of energy through conservation measures and improved efficiency. Nigel Mortimer came up with a similar conclusion for the United Kingdom. He found that energy conservation strategies, such as the use of low energy lighting, cavity wall insulation or loft insulation would be extremely cost effective in saving on the need to generate electricity.

A commitment to alternative energies requires governments to make fundamental changes to the way they operate their economies. At present all manner of subsidies and interventions favour the development and use of certain forms of energy at the expense of others. Private transport tends to be favoured over public transport through pandering to a powerful road lobby; nuclear power has received massive government support in the form of hidden subsidies, some linked to the development of nuclear arsenals; coal mining has been heavily subsidized in the past.

No environmental damage, including acid rain and habitat loss, damage to health, climatic effects, has registered as a cost to be added to the market price of coal, oil, or indeed nuclear power. The biggest subsidy has been in the failure to account in any way for the exploitation and destruction of natural resources.

If governments practised what they preached in advocating a competitive market economy, and natural resources were properly valued as *capital* rather than as *income*, such a form of economic valuation might go some of the way to realizing the true worth of many alternative energy systems, not least those which lead to improvements in energy efficiency and conservation. A study carried out by the German Bundestag in 1990 showed that if the social and environmental costs of energy generation are taken into account, then wind power is twice as competitive as conventional electricity generation from coal.

Economists talk of *no regrets measures* that can be undertaken to mitigate the effects of global warming. The point is that such measures make economic sense irrespective of global warming. The application of a market system that recognized social and environmental costs as intrinsic costs rather than external costs and properly accounted for the user cost of natural resources, would enhance considerably the use of all manner of *no regret* measures. Were such measures applied, few would be able to afford the full costs of urban driving and cities would not need to ban private vehicles from their centres. Similarly, utilities would not need to seek subsidies for renewable energy systems; on the contrary they would not be able to afford highly polluting fossil fuel systems.

As Christopher Flavin of Worldwatch points out, alternative energy systems are already beginning to penetrate the market, even without applying *fair* user cost criteria. Advanced electronics have improved the efficiency of lighting fourfold and wind power has more than doubled its contribution to electricity production from 2000 MW in 1990 to 4500 MW in 1995. Solar energy has also become a mature technology and in 1993 in Kenya more homes were electrified using solar cells than were connected to a central grid system. Japan too has undertaken a scheme to install 62,000 solar systems that are integrated into the structure of new buildings by the end of the century. Greece too has recently installed the largest photovoltaic electricity generating system in the world with a generating capacity of 10 MW. The cost of solar cells has come down from $70 per watt in the 1970s to $4 per watt today and is likely to cost no more than one or two dollars within a few years. Solar cells on rooftops could provide as much as one quarter of the world's electricity by 2050.

When we include other technologies like fuel cells that use fossil fuels more cleanly, or flywheels that conserve energy, we begin to glimpse a future that makes better use of resources than we do today. Nevertheless, China is calling on help from US manufacturers to build massive power plants based on coal burning technology that, at best, is only forty per cent efficient. In its drive to expand its economy, China clearly cannot wait for improved energy systems.

Chapter Thirteen: The Impact of Climate Change

MODERN technology has given us the power to rampage across the planet. No ecosystem has been left untouched and many have been disastrously degraded. Moreover, the modern way of life has usurped other human cultures thereby destroying ways of life that have been remarkably adapted to specific environments. Yet, for all our modern knowledge and technologies, we have remained woefully ignorant of the impact of what we are doing to the planet. We have, for instance, largely failed to realize that our climate and therefore weather, is a consequence of the multitudinous activities of living organisms, from the lowliest of bacteria to the massive sperm whale. They all play a role in the cycles of life and the interchange of gases and materials across the interface between inanimate rocks and the atmosphere.

The idea of Gaia is of profound significance as a dynamic, self-regulating system that helps stabilize climate against geophysical perturbations. A stable climate is therefore synonymous with a healthy environment of interacting ecosystems. If we are busily engaged in the systematic destruction of those ecosystems, in the name of development and progress, what will such vandalism do to climate?

Climate Models

The problem is that we simply do not know how resilient the total system of life and climate is. Our climate models are relatively simplistic and do not incorporate the notion of self-regulating cycles bound by certain limits. What are those limits? Have we already exceeded them, or are we in danger of doing so? We may never be able to prove the concept of Gaia as a self-regulating planetary system. But, now that we are finding evidence of planetary scale phenomena, such as the biotic regulation of atmospheric oxygen, we would be foolish to dismiss the idea of Gaia and the possibility that its healthy functioning is under threat.

Unfortunately, we have tended to treat the environment like a sponge with a never-ending capacity to absorb whatever we might throw at it. But sponges have their limits and suddenly can absorb no more. This analogy is crucial for understanding what happens in the soil, water bodies and the atmosphere, as we continue to discharge our industrial effluents, engage in intensive, high input agriculture and support high energy lifestyles. Natural systems have a certain capacity to protect themselves against change, and all might appear well to the onlooker right up to the point of saturation: then the system fails. We must take the possibility of failure into account when it comes to predicting climate change, and must critically question the assumption of climate models that the basic system of greenhouse gas interchange remains intact and functioning, when in reality it may have broken down irreversibly.

Acid Rain

We can obtain a disturbing foresight of what might happen to climate systems by following the history of the impact of acid rain on soils, forests, rivers and lakes. Rain is naturally somewhat acid from the dissolved carbon dioxide, as well as the dissolved oxides of nitrogen and sulphur. The flow of sulphur and nitrogen compounds from land to ocean to atmosphere are part of natural cycles. Fossil fuels contain considerable quantities of sulphur, and release sulphur dioxide into the atmosphere when burnt. Combustion also produces nitrogen oxides from the nitrogen in the air. When the sources of sulphur dioxide and nitrogen oxides are taken into account, then the quantities amount to as much, if not more, than all the natural sources of those substances put together. Mixed with water, the oxides become acids.

The nickel and copper smelting plant at Sudbury in Ontario is the largest man-made emitter of sulphur dioxide in the world. Not surprisingly Canadian rivers and lakes have suffered as a result and 300 lakes in Ontario have pH values below five, and can no longer support fish. The acidification of the soils is also causing sugar maples to die-back.

Since the end of the last century Central Europe has been home to heavy industry, with factories belching out sulphur-laden smoke. In the heart of Poland's industrial wasteland, acid rain has eaten at railway lines, destroyed the façades of historic buildings in the ancient town of Crakow, and generated a host of respiratory problems.

The soils of Central Europe have been subject to at least a ten times more sulphur over the years than southern Sweden, and soil scientists are now concerned that soils are at the point of saturation.

Acidic Soils

Lake Blamissusjon in northern Sweden exemplifies what can happen when soils dry out. The lake originally emerged from the sea after the last ice age. Its bottom sediments were of marine origin and rich in insoluble sulphides. The surrounding soils were waterlogged and the sulphide remained in its reduced, non-oxidized, state. When the land was drained for agriculture, in 1900 and more extensively in the 1940s, the sulphide was exposed and oxidized to sulphuric acid. In a matter of years, its pH dropped to three, making it the most acidic lake in Sweden. Even though farming in the region was abandoned by 1970, the pH levels remain as low as ever.

Soils as Buffers

In their capacity to neutralize acids, soils in watershed areas help protect rivers and lakes from acid rain, but soils overlying low weathering rocks such as granites, sandstones and quartzites, tend to be naturally acidic, and their ability to absorb more acid is strictly limited. Consequently, certain regions, such as the Adirondack Mountains of the United States, or the igneous bedrock areas of southern Sweden, are particularly vulnerable to the effects of acid rain that has been enhanced by acid fall-out from industrial activities. For a time the natural system copes and all appears to be well, but with its buffering capacity exceeded, the system changes, causing the ecosystem to collapse. Because the response is often delayed and we do not see any immediate effects, we get a false sense that all is well right up until the time the system crashes.

Acidification of European rivers and lakes first came to light in Norway in the 1920s, with a noticeable decline in the salmon population. Lakes and rivers in Sweden have also been affected to the point when not even the most acid-resistant fish can survive. Out of Sweden's 90,000 lakes, 20,000 are now acidified to some extent and 4,000 are devoid of fish. In Norway, the situation is even worse, with fish having been exterminated from lakes covering 13,000 km². Other studies show that soils in deciduous forests of southern Sweden have become progressively more acidic since 1949, with the result that aluminium, a toxic metal, has begun to leach out of the clays and to replace calcium and potassium. The aluminium is responsible for fish death and is implicated in forest die-back.

Big Moose Lake in the Adirondack Mountains of New York State appeared to be fine until the 1950s then, over the space of 30 years, the acidity went up fifteenfold. The acid came from sulphur dioxide emitted from the operation of coal-fired plants in the Ohio Valley since the 1880s.

The acid rain problem was compounded by the policy established in the 1960s to discharge flue gases through tall chimney stacks, in the belief that the gases would disperse in the atmosphere. Unexpectedly, the plume remained relatively intact over hundreds of kilometres, so that particulates and gases either washed out in rain or falling *dry*, landed more or less over the same location. Because of the prevailing winds, the flue gases from power stations in the United Kingdom and from Central Europe, tended to fall-out over Scandinavia, which therefore received far more than its fair share. Only one fifth of the sulphur dioxide fall-out over Sweden was of local origin. The discharge of the gases several hundred metres into the atmosphere also had another unanticipated effect; intense sunlight converted the chemicals to a lethal mix of corroding acids and organic matter.

It is no coincidence that the worst affected areas of tree death and die-back in Germany's Black Forest lie at the same altitude as the gas emissions from chemical factories 300–400 m below in the Rhine Valley. In the winter, the plumes drift in intense sunlight just above the lower inversion layer. Then, at night, as the temperature falls below freezing, the mix of chemicals is caught in ice droplets that form on the branches of the fir trees.

Ultraviolet Penetration

As the waters become acid they tend to become increasingly translucent. Scientists are now concerned that acidified lakes are facing an additional hazard through increased ultraviolet penetration. In Eastern Canada, out of 700,000 lakes, the surface waters of one 140,000 are now so clear that eight times more ultraviolet is getting through the surface waters. As a result, algal communities with bright protective pigments have begun to take over from light sensitive species.

As David Schindler, of the University of Alberta, has discovered,

the ecological changes in Canadian waters are the result of a combination of factors, including global warming, and not just acidification alone. He has been studying the ecology of the lakes for more
than twenty years, over which time he has seen a precipitous decline
in the amount of feedwater running into the lakes from streams and
rivers. The waters carry dissolved organic matter which helps feed
algae and other living organisms, but the dissolved matter also
clouds the surface waters so that ultraviolet light, especially UV–B,
is unable to penetrate.

Over the past two decades average surface temperatures in the region have gone up by 1.6°C. Concurrently, rainfall has dropped by
one quarter, while evaporation has increased by one third. Streams
that flowed all year round are now dry for nearly half the year. Even
lakes that are broadly unaffected by acid rain have become clearer
because of the drop in feedwater carrying organic matter. UV–B
penetration in these waters has increased by at least sixty per cent.

Transboundary Pollution

In 1979, under the auspices of the UN Economic Commission for
Europe, 33 countries in Europe and North America signed a convention on *long-range transboundary air pollution,* the aim being to
reduce both sulphur dioxide and nitrogen oxides by at least thirty per
cent compared with 1970 levels. Since most of the sulphur dioxide
is generated by fossil fuel burning power stations, on-site technologies can be applied to capture emissions and prevent contamination
of the air.

Sulphur Reductions

By 1988 the United States had achieved a 20% reduction in sulphur
emissions, and Britain nearly 50%. France, after an initial rise in
emissions in the early 1980s, cut back to 40% of what they produced
in 1970, West Germany 67%, Sweden 60%, while Japan and the
Netherlands reduced them to 20% of what they were thirty years before. The economic and political collapse of the Soviet Union and
the European Eastern Bloc also led to major reductions in emissions.
By 1992, despite cutbacks in the United States and Europe, a 50%
rise in worldwide emissions of nitrogen oxides, and 20% in sulphur
dioxide, took place .

Chinese Emissions

By comparison with other nations, both developed and developing, China has an extremely inefficient economy. The concern is, that as China's economy grows, coal burning is likely to increase commensurately. Some forecasts predict that China will burn more coal than all the developed countries of the OECD (Organization for Economic Cooperation and Development) by the turn of the century. The atmospheric consequences of burning such a vast quantity of coal will be catastrophic unless combined with the latest technologies to improve efficiency, and control sulphur and nitrogen.

Paradoxically, in Europe and the United States, where sulphur precipitation has declined substantially since the late 1970s, soils have shown little or no improvement. One possibility is that the buffer capacity has been totally taken up from the fall-out of previous years. Another reason is that the emissions of acid precursors were accompanied by emissions of calcium, magnesium, potassium and sodium, all of which as *base cations* help neutralize the effects of acid. However, they have also been cleaned out of the stacks, together with sulphur and nitrogen oxides.

The clean up of sulphur emissions over North America and Europe has had other effects; Clearer skies, with less dust, smog and cloud may be tipping the balance towards global warming, since more sunshine penetrates the Earth's surface.

Lack of Sulphur

Sulphur is an essential nutrient and as long as soils have been kept limed, farms have benefited from the fall-out of fossil fuel sulphur. In the late 1970s 50 kg per hectare of sulphur fell out each year over European farmland. Today, because of clean air programmes, sulphur fallout has fallen to less than 10 kg per hectare. New crop diseases, especially in members of the brassica family, such as oilseed rape, have started manifesting themselves. Lack of sulphur causes the leaves to thicken and turn yellow and the few, faded flowers that are produced, repel rather than attract bees. The plants become susceptible to fungal diseases and their response to nitrogen fertilizer is poor, thereby exacerbating nitrate run-off into groundwater and leading to more nitrous oxide reaching the atmosphere.

According to Ewald Schnug, at the Institute of Plant Nutrition and Soil Science in Brunswick, Germany, more than a century ago farmers started applying chemical fertilizers in the form of ammonium sulphate. After World War II, farmers switched to using ammonium nitrate and sulphur-free forms of phosphate fertilizer. But that was when economic activity boomed and sulphur fallout reached its maximum. Schnug believes that sulphur acts in the plant to counteract the effects of oxidizing chemicals, such as ozone. Ozone levels close to the ground have increased as a result of pollution from motor vehicles.

The ways in which soils and sediments react to drying out, and to changes in oxidation as well as acidity, can have profound effects on the solubility and availability of chemical deposits, in particular toxic chemicals that have been discharged or disposed of by industry and agriculture. Changes in land use, perhaps even as a result of global warming, could unleash a chemical time bomb.

Many farmers offset soil acidification by spreading lime. However, with set-aside programmes, they no longer lime the land and the pH may plummet. If that land has been treated with a herbicide such as paraquat, which may be retained on clay particles for fifty years or more, acidification could cause the cations in the clay to become soluble, with the result that paraquat will flush into groundwater, and perhaps into drinking water supplies.

Toxic Flush

Each year more than 10,000 tonnes of toxic chemicals flush from the mouth of the Rhine into the North Sea. Over a century the accumulation of such discharges has left sediments rich in highly toxic heavy metals and other substances. To keep navigation lanes open around Rotterdam about fifty million tonnes a year of sediment is dredged out. Until the 1980s the sediment was used to fill polder land in Holland, as part of the scheme to make available more agricultural land while keeping the sea out. The dredging of the sediment, its exposure to air and the increase in acidity has led to the mobilization of toxic metals such as arsenic, cadmium, chromium, copper, mercury and lead, as well as organic organochlorines. Crops have become contaminated and the practice of using the sediment is now banned.

Wetlands as Sinks

Wetlands serve as sinks for sulphur compounds and under anoxic conditions bacteria in the soil rapidly convert sulphate to immobile sulphide compounds such as iron pyrite (see Chapter 7). Wetlands therefore improve nearby farmland by reducing its potential acidity. But what happens if wetlands dry out because of global warming? Sulphide oxidizes, thereby increasing the acidity of the surroundings. Denitrification, by which nitrates are reduced into nitrogen gas, also fails when wetlands dry out, and nitrate can now run off contaminating ground and surface waters.

Global warming models suggest that once carbon dioxide levels in the atmosphere have doubled, soils in southern Europe may lose fifty per cent of their moisture, and soils in northern Europe thirty per cent. Some soil scientists fear that the pH of European soils will plunge down to four or less. If so, calcium and other basic cations will leach out, and will be increasingly replaced by aluminium, which itself will become more soluble, with consequences for vegetation and aquatic life.

Increasing soil temperatures have a marked effect on the capacity of soils to retain organic matter. William Stigliani, of the International Institute for Applied System Analysis in Austria, points out that if the annual mean temperature of soils increases by 10°C, and soil moisture falls by thirty per cent, levels of organic nitrogen in the soil will drop by nearly half. The decrease in organic matter has a strong effect on the capacity of the soil to retain important nutrients for plant growth. Drier soils, as we have seen, are likely to be better aerated so that soil-held wastes are oxidized, and no longer bound tightly to constituents of the soil. Dry soils are also more likely to blow away and erosion rates are bound to increase.

Pesticides, heavy metals and other toxic substances currently held secure in organically-rich soils would suddenly be mobilized. A toxic catastrophe could be unleashed over large areas of farmland, particularly those that have been intensively managed.

Coral Reefs

Virtually all ecosystems are now under seige, not least the coral reefs of the world. In 1986, at the height of the monsoon, storm surges battered at Sri Lanka's coastline, tearing up beaches, demolishing houses and washing away roads railway lines. Most of the damage was along sections of the coast where the protective ring of coral reef had gone, destroyed to make concrete for the country's housing boom.

Covering 600,000 km² coral reefs are to be found in the warm, shallow waters of the Tropics. They comprise some of the most species-diverse ecosystems in the world, and have been particularly badly hit in recent years. Mining their lime is not the only cause of their destruction. In many parts of the world, fishermen are using dynamite to fish, irrespective of the damage done to the reef.

In the Tropics the recent destruction of rainforests to make way for farmland and cattle ranching has led to a significant increase in the amount of soil being washed away and carried down to the ocean. Corals are being smothered and as algal blooms take over and blot out the light, they are dying. Logging in the watershed of Bascuit Bay in the Philippines has doubled the amount of sediment being washed downstream. At least five per cent of the coral reefs in the Bay have died, and the remainder is in a parlous state. If the rates of decline and destruction to corals continue, then within forty years as much as 75 per cent may have vanished for good, taking 200,000 species of animals with them.

Coral Bleaching

Global warming is another hazard, since the photosynthesizing *zooxanthellae* that live symbiotically with the coral, are unable to live in waters with temperatures above 30°C. In 1987 and 1989 fishermen in the Caribbean discovered that the corals were *bleaching*. Coral reefs are therefore under threat just when we are likely to experience a significant increase in tropical storms and cyclones, also brought about by warming.

Apart from the protection they provide to vulnerable, low-lying coastal lands, coral reefs are important nurseries for valuable fish and shellfish. They are also an integral part of the capture of carbon dioxide and its transformation into limestone, and therefore make a valuable contribution to the stabilizing of greenhouse gases. Conceivably too, by forming a natural barrier over geological periods of time and enclosing areas of ocean, which leave the salts behind through evaporation, corals also play a role in the regulation of salt concentration in the oceans.

Warming Feedbacks

The latest reports from the Hadley Centre of the UK Met Office state clearly that worldwide we are sliding inexorably into a period of global warming. According to its analyses, within fifty years we will face climatic disturbances that could even put our future at risk. When we begin to account for some of the essential positive feedbacks, in particular those involving life, we begin to see how much more dangerous the situation is than anything we may have envisaged.

In testing their general circulation models, climatologists try to get as close correspondence as they can against data of surface temperatures over the past 140 years. By averaging out temperatures over a succession of 30-year periods they obtain a *reference* temperature for each year. Once relatively good correspondence is achieved, they feed special parameters into their models, such as rising CO_2 levels, to accord with different scenarios of greenhouse gas emissions. The intention is to obtain predictions of sea level rise, air circulation systems, precipitation and not least surface temperatures in different parts of the globe.

However, the parameters used to establish concordance with the past may have proved adequate until now, but the likelihood is that they will prove deficient in the future, since complex feedbacks are now coming into play that have not yet been modelled. For example, the impact of ocean warming on populations of phytoplankton, which in turn will effect carbon dioxide absorption. In fact, from 1860 until World War II changes to the global environment were limited, but over the past fifty years changes to the environment have been radical and processes such as forest die-back and desertification have become self-reinforcing. In this respect, the recent past will be no guide to the future.

Impact of Warming on the Ecosphere

Our concern is largely with positive feedbacks, triggered in this instance by global warming and other impacts of economic development. Changes to the environment, like rising temperatures and the destruction of natural ecosystems, may become amplified in a feedback process that spirals out of control, producing runaway warming.

Another complicating feature of positive feedbacks is their tendency for distinctly different processes to feed on each other in synergistic interactions. For instance, global warming causes soils to dry out which in turn leads to the die-back of forests. The die-back of forests leads to the further drying out of soils, as well as to the release of carbon dioxide and methane. The additional greenhouse gases add to warming and hence to further deforestation. Meanwhile, surface winds increase over the deforested areas because of greater contrasts between day and night temperatures. The soil, now dry and decomposed, starts blowing away.

Vegetation and the Effects of Global Warming

Richard Betts from the Hadley Centre, and his colleagues from Sheffield University, have tried to measure what will happen to vegetation as a result of increases in atmospheric carbon dioxide and climate change. They get mixed results. More carbon dioxide means better growth and more efficient use of water, but that may have drawbacks, especially in the Tropics where evapotranspiration is an important mechanism for cooling and for feeding rain clouds so that the forest remains well watered.

Meanwhile, more vigorous growth in high latitude, boreal regions, can accentuate warming by bringing about earlier snow melts, thus exposing the leaf-darkened surface to the Sun. A conifer, for example, is aptly shaped to shed snow, thus exposing the dark, green needles to the first rays of the spring Sun. The great boreal forests therefore bring winter to an end much faster than other vegetation would, while also extending the summer. Should global warming cause the northwards spread of conifer forests that will bring about more warming.

Global warming could be pushing the climate towards a regime that favours boreal forests at the expense of *Sphagnum moss*. If so, rising temperatures, from the melting of permafrost, could bring about the atmospheric release of as much as 450 billion tonnes of carbon in the form of carbon dioxide and methane (see Chapter 3).

Terrestrial Sinks and Sources

On the basis of a steady increase in carbon dioxide, the UK Met Office's Hadley Centre predicts a sharp increase in the mass of tropical forests over the next fifty years. That sounds like good news, but the entire tropical ecosystem would then collapse abruptly. The forests, which currently absorb about one third of the carbon dioxide we emit, will no longer be able to cope with the drastic reduction in rainfall, combined with temperature rises as high as 8°C. Not only will the forests decompose as they die-back, but their carbon dioxide absorbing mechanism will no longer function and we will be left with a surge in greenhouse gases. Tropical forests contain approximately forty per cent of the carbon contained in terrestrial biomass, which amounts to 550 billion tonnes in total. The release of most of the carbon now contained in the world's tropical forests would be equivalent to one third the carbon in the atmosphere.

We might be misled into thinking that the modellers have shown us the worst scenarios. Yet, by their own admission 'the model describes the potential natural vegetation that would exist without interference by humans, such as the recent rainforest fires.' The actual situation is therefore fundamentally different from the one they are modelling.

In fifty years, we have lost approximately half of the world's tropical forests and, at the rate we are going it will not be long — possibly as little as 30 years — before we have lost the rest. Under these circumstances can we take the Met Office's prediction seriously that a surge in tropical forest growth over the next fifty years will remove an extra billion tonnes of carbon a year from the atmosphere, taking the total annual terrestrial carbon sink from two to three billion tonnes? If this extra sink is not there, because of *premature* forest destruction, the terrestrial sink will have turned into an emission source.

Global warming could also be on the verge of triggering the destruction of the West Antarctic ice sheet, which in its entirety could cause a worldwide rise in sea levels more than ten times the current predictions. (see Chapter 6). The Greenland Ice Sheet, grounded as it is on land, is far more stable than the West Antarctic Ice Sheet, but were it to melt because of rising temperatures, it would add another six metres to sea level rise. As the Met Office reminds us, global warming will be at its most intense over the Arctic. By its predictions, if greenhouse gases increase at their current rate of one per cent a year over the next fifty years, some regions in the Arctic Circle would warm by as much as 6°C compared with today — two to three times greater than the global average.

According to the US Geographical Survey, 10,000 billion tonnes of methane are currently trapped under pressure in crystal structure on the edges of continental shelves, making them the Earth's largest fossil fuel reservoir. If the temperature in the surrounding water or sediment is increased to the point where a methane hydrate becomes unstable, methane gas would be released overnight. Where water is relatively shallow and easier to heat, as in the Arctic, tens if not hundreds of billions of tonnes of methane could be released.

Losing the Ocean Sinks

The continued function of the oceans as a net sink for atmospheric carbon dioxide is of critical importance for a stable climate. If this sink is lost, then that might unleash another positive feedback which the climate models have yet to take fully into account. The circulation of the oceans is vital in both the uptake of carbon dioxide from the atmosphere and the transport of heat from the Equator to the high latitudes (see Chapter 6). A stalling of the conveyor belt will have repercussions for the ocean's ability to absorb anthropogenic carbon dioxide from the atmosphere. Two billion tonnes of carbon are at stake — between one third and one quarter of the carbon dioxide emitted from fossil fuel and forest destruction.

If the population of phytoplankton begins to crash because of a combination of warmer surface waters, a curtailing of the conveyor belt, and exposure to ultraviolet from the ozone hole, then much less carbon dioxide will be drawn down into the ocean depths: hence more warming.

Even without taking changes in biological activity into account, Jorge Sarmiento and Corinne le Quéré at Princeton University, find from their models that both stratification and a slowing down of the conveyor belt will lead to as much as a fifty per cent reduction in carbon dioxide absorption into the ocean. A warming of the oceans by an average 5°C as a result of the quadrupling of carbon dioxide in the atmosphere would lead over time to an additional 50 ppmv of carbon dioxide — one seventh of current levels — in the atmosphere.

Not only are we on the threshold of losing an annual two billion tonnes sink for carbon dioxide but, as the oceans warm the solubility of carbon dioxide drops significantly. We will therefore begin to see the additional carbon dioxide bubbling out of the surface waters. Instead of a vital sink for carbon dioxide the oceans could turn into a net source. With fifty times more carbon dioxide in the oceans than

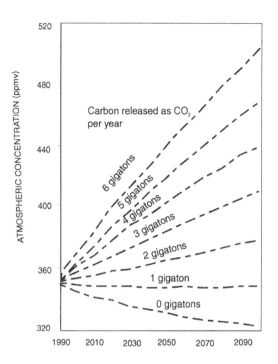

Figure 27. Atmospheric concentration of carbon dioxide which contributes to global warming. (Souce: Scientific American, *September 1990).*

in the atmosphere, it takes no more than a subtle change in the exchange between the oceans and atmosphere for greenhouse gas levels to double.

Clouds and Global Warming

The cloud forming attribute of phytoplankton is of vital importance for climate. Fewer clouds means a warmer ocean which means less phytoplankton and less clouds. When we add this feedback to that of oceanic carbon dioxide uptake, we have a formidable combination on our hands that could transform global climate in a matter of years.

Meanwhile, terrestrial vegetation in the Tropics and in the boreal regions have powerful climatic impacts, particularly as a result of their effect on albedo and energy transfers. The energy pumped into the atmosphere by the Amazon rainforest is practically equivalent to the energy transported northwards by the Gulf Stream. Not only are the rainforests of the Tropics powerful pumps but they simultaneously release cloud condensation nuclei. That cloud forming, energy transporting system is extraordinarily vulnerable because of its dependence on adequate watering. If the pattern of watering is broken, for instance by a succession of El Niños, then the Amazon rainforest is doomed — that at least emerges from the Hadley Centre models.

The activity of life in the two domains, the oceans and the continents, in terms of cloud formation and the hydrological cycle, is therefore of crucial importance to climate. Yet, such feedbacks are not currently integrated into climatologists' general circulation models.

Living on a Knife-Edge

The point is disturbingly clear. The general circulation models, although vastly improved in recent years and based on excellent science, do not come close to evaluating the true impact of our activities on global climate. Life is the key and we must incorporate feedbacks if we are to come near to grasping the consequences of our rampage across the planet. The issue is not just greenhouse gases and global warming — although these issues are undoubtedly crucial — it is also our destruction of vital ecosystems. We call this destruction *progress*. We build dams to harness water for irrigation and hydro-electricity, so setting in motion dramatic changes to oceanic circulation because of a decline in the flow of freshwater. We are methodically compacting, eroding, salinizing and desertifying our agricultural land. For every hectare of *improved* land for modern intensive agriculture we leave another behind that has become desert. We are grubbing out coral reefs for cement works and road building; draining our wetlands to accommodate yet more export oriented monoculture; and we have already destroyed more than half the world's tropical forests, with little to stop the rest going within the first decades of the coming century.

The Consequences of Climate Change

We now face the danger that many interconnected, though separate positive feedbacks could be triggered at the same time, all acting synergistically to exaggerate the impacts of the other. It is most unlikely that we have identified all of them. Currently the atmosphere contains 750 billion tonnes of carbon, equivalent to 365 ppmv of carbon dioxide. If 450 billion tonnes of the carbon from the wetlands were added to the levels already there, it would take the total atmospheric carbon content to over 515 ppmv. Add the collapse of terrestrial ecosystems such as tropical forests, thus turning the land surface into a net source of carbon, and over one hundred years we would have at least an extra 200 billion tonnes of carbon in the atmosphere — approximately equivalent to an extra 100 ppmv.

Imagine too, that we continue to emit at least 6 billion tonnes per year of carbon from fossil fuel burning, (it has been up to 6.3 billion tonnes) and one hundred years from now, from that source alone, we would see a tripling of pre-industrial levels of carbon dioxide. In addition, we are still relying on the oceans to draw down some two billion tonnes per year of fossil fuel carbon emissions. What if the ocean carbon sink also collapses? In one hundred years' time, rather than the doubling of carbon dioxide anticipated by climatologists, we will see a quadrupling, with unthinkable consequences to climate.

A fourfold increase in the greenhouse gases from pre-industrial times appears inevitable if we carry on as we are. That will take the levels of carbon dioxide even beyond 1100 ppmv. That would be the highest the levels have been for more than one hundred million years. And, we know from ice cores that the global climate is perfectly capable of making dramatic shifts in a decade or less.

We have undoubtedly reached a watershed. We can either continue as we are and hope that we will be able to contain the damage, not least from climate change, or we can decide to do everything in our power to reduce our impact on the environment before it is too late. Certainly climate change is teaching us some crucial lessons. One is that we cannot predict the future of climate, we can only guess at it. That conclusion makes it glaringly apparent that we should disabuse ourselves of any notions that we can manage climate.

Global decisions to modify human behaviour because of the threat of climate change, even under the auspices of the United Nations, will never be truly effective while the Earth's natural resources are handed over to corporate, profit-seeking interests. Management of climate will emerge quite naturally from a world in which communities, rather than states or corporations, become responsible again for their lands and for seeing that they are maintained in a good state. We need to learn how to manage ourselves and our relationship with our environment. Then and only then might we have some hope that we will be able to bring to an end the destruction that is threatening our existence.

Appendix I

IPCC Scenarios of Energy Use

Taken from Climate Change: The IPCC Scientific Assessment, *Cambridge University Press, 1990.*

The following scenarios cover the emissions of carbon dioxide, methane, nitrous oxide, chlorofluorocarbons, carbon monoxide and nitrogen oxides, from the present up to the year 2100. Economic growth and population was taken as common for all scenarios. Population was assumed to reach 10.5 billion in the second half of the next century. Economic growth was assumed to be two to three per cent annually in the coming decade in the OECD countries, and three to five per cent in Eastern European and deveopling countries. Economic growth was seen to decrease thereafter. In order to reach the required targets, levels of technological development and environmental controls were varied.

Scenario A — the high growth, business-as-usual scenario — sees the world economy growing at 3.8 per cent per annum, made up of a 2.4 per cent growth rate in the developed world and a 5.6 per cent in developing countries. Energy supply is coal intensive and only modest efficiency increases are achieved on the demand side. Carbon monoxide controls are modest, deforestation continues until the tropical forests are depleted,and agricultural emissions of methane and nitrous oxide are uncontrolled. The Montreal Protocol is only partially implemented with regard to CFCs. Even with a 1.6 per cent per year improvement in the use of energy per unit production, by 2020 energy demand is forecast at 23.64 TW. By 2100, energy demand has leapt to 57.7 TW, with annual carbon emissions at 16.6 Gt, three times higher than today. Thirty per cent of primary energy is from nuclear power, with fossil fuels providing forty per cent, and renewable energies the remainder. The nuclear power commitment would be 28 times greater than it was in 1990, or 16.73 TW compared to 0.6 TW — 16,000 large nuclear power plants compared to 600 today.

Scenario B has more modest aspirations: 3.3 per cent growth per annum as a worldwide average. The energy supply mix shifts towards lower carbon fuels, notably natural gas. Large efficiency increases are achieved. Carbon monoxide controls are stringent, deforestation is reversed and the Montreal Protocol implemented with full participation. The WEC has come up with two alternatives: *Scenario B1*, in which the reduction in energy use per GNP is 1.3 per cent per annum and *Scenario B,* in which the reduction is 1.9 per cent. The difference is seen in the energy demand which is 22 TW compared to 18.4 TW. By the year 2100, Scenario B gives an energy demand of 45 TW with annual carbon emissions nearly double today's and nuclear power providing 12.7 TW equivalent to meeting all of the world's current energy needs.

Both Scenarios A and B raise real doubts as to whether such energy demands can be met; and if they can, whether the world as we know it would have any chance of surviving. The thought of so many nuclear power plants, with their requirement for cheap uranium, is quite daunting. The amount of plutonium generated each year in the operation of such plants would lie between 4 and 5.5 thousand tonnes a year — enough for hundreds of thousands of atomic bombs.

Scenario C is based on a high reduction in energy intensity of 2.4 per cent per year and a modest increase in worldwide energy use by the year 2020 to 15.5 TW. There is a shift towards renewable energies and nuclear energy in the second half of the century. CFCs are now phased out and agricultural emissions are limited. By the year 2100, energy use will more than double its 1990 level to 27.5 TW, but by reductions in the fossil fuel burn, carbon emissions will be under half their current level, at 2.5 billion tonnes per year. In this scenario half the energy comes from renewable sources and eleven per cent from nuclear power. Nuclear power would therefore be providing five times more primary energy worldwide compared with today. Some argue that nuclear power is too dangerous a technology to use at all, and that energy demands must be tailored to meet more modest needs than those proposed in Scenario C.

Scenario D involves a shift towards renewables and nuclear energy in the first half of the century. The scenario shows that stringent controls in industrialized countries combined with moderate growth of emissions in develpoing countries could stabilize atmospheric concentrations. Carbon dioxide emissions are reduced to fifty per cent of 1985 levels by the middle of next century. However, the dependence on nuclear power raises valid concerns about safety.

Appendix II

Ozone Depletion Potential

On applying the ODP, solvents such as carbon tetrachloride (CCl_4) and methyl chloroform ($C_2H_3Cl_3$) contribute 7.6 per cent and 5.1 per cent respectively to ozone depletion, against CFC-11's contribution of 30.4 per cent, CFC-12's of 40 per cent and of CFC-113's of 11.7 per cent. The halons and CFC-114 contribute 5.2 per cent to the total.

Appendix III

Table of Geological Periods

Geological Periods	Years before Present
Younger Dryas	12,000
Eemian	20,000
K/T Boundary	65 million
Cretaceous	136 million
Permian	280 million
Carboniferous	345 million
Devonian	395 million
Ordovician	500 million
Cambrian	570 million
Pre-Cambrian	pre-570 million
Proterozoic Era	2500–1900 million
Archaean Era	4600–1900 million
Hadean Era	pre-Archean Era

Glossary

AOSIS	Alliance of Small Island States
billion	10^9 (one thousand million)
CCN	cloud condensation nucleii
CFC	chlorofluorocarbon
DMS	dimethyl sulphide
ENSO	El Niño Southern Oscillation
EPA	Environment Protection Agency
FAO	UN Food and Agriculture Organization
GCI	Global Commons Institute
GCM	General Circulation Model
GDP	Gross Domestic Product
GNP	Gross National Product
Gt	gigatonnes (10^9 tonnes)
GtC	gigatonnes of carbon (billions of tonnes of carbon)
GWP	Gross World Product
HCFC	hydrochlorofluorocarbon
HFC	hydrofluorocarbon
IPCC	Intergovernmental Panel on Climate Change
ITCZ	Intertropical Convergent Zone
Met Office	UK Meteorological Office
1μ	1 micrometre (10^{-6} micron)
MW	megawatts (10^6 watts)
NASA	National Aeronautics Space Agency
NRPB	National Radiological Protection Board
ODP	Ozone Depletion Potential
ppmv	parts per million by volume
trillion	10^{12} (one million million)
TW	terawatts (10^{12} watts)
UNCED	UN Conference on Environment & Development
UNEP	UN Environment Programme
UV–A	ultraviolet A
UV–B	ultraviolet B
UV–C	ultraviolet C
WEC	World Energy Council
Wm^2	watts per square metre

Bibliography

Boyle, Geoffrey (Ed), *Renewable Energy: Power for a sustainable Future,* Oxford University Press, 1996.,

Brown, Lester R., *Facing Food Scarcity,* Worldwatch, Washington DC, Nov/Dec. 1995.

Brown, Lester R., *Tough Choices: Facing the Challenge of Food Scarcity.* Norton, New York, 1996.

Bunyard. Peter (Ed), Gaia in Action: Science of the Living Earth, Floris Books, 1996.

Bunyard, Peter, 'Eradicating the Amazon Rainforests will wreak havoc on Climate,' The *Ecologist,* Vol. 29, No. 2, 1999.

Daly, Hermann E. and Cobb. John B., *For the Common Good,* Beacon Press, Boston, 1989.

Department of the Environment, Scottish Office, *The United Kingdom National Air Quality Strategy,* HMSO, 1997.

Downing, Thomas E., Olsthoorn, Alexander J. and Tol, Richard S J, *Climate Change and Risk,* Routledge, 1999.

Eddy, John, 'Solar History and Human Affairs,' *Human Ecology,* Vol. 22, No. l, 1994.

Gelbspan. Ross. *The Heat is On,* Addison Wesley, Reading, Massachussets, 1997.

Goldsmith, Edward, *The Way: An Ecological World-View,* University of Georgia Press, 1998.

Grace. John. 'Forests and the Global Carbon Cycle,' *S. It. E. Atti,* 17:7-11, 1996.

Handler, Paul and Andsager, Karen, 'El Niño, Volcanism and Global Climate,' *Human Ecology,* Vol. 22, No. 1, 1994.

Ho, Mae-Wan, *The Rainbow and the Worm: The Physics of Organisms,* World Scientific, Singapore, 1998.

Houghton, John, *Global Warming: The Complete Briefing,* Lion Publishing, Oxforf, 1994.

Houghton, J.T., Jenkins, G.J. and Ephraums, J.J. (Eds), *Climate Change: The IPCC Scientific Assessment,* Cambridge University Press, 1990.

Intergovernmental Panel on Climate Change, *1992 IPCC Supplement: Scientific Assessment of Climate Change,* 1992.

Intergovernmental Panel on Climate Change, *Climate Change 1995: The Science of Climate Change,* UNEP, 1995.

Jepma, Catrinus J. and Munasinghe, Mohan, *Climate Change Policy,* Cambridge University Press, 1998.

Joint Global Ocean Flux Study, *Oceans, Carbon and Climate Change,* 1990.

Lamb, H. H, *Climate History and the modern World,* Methuen, 1982.

Leakey, Richard and Lewin, Roger, *The Sixth Extinction: Biodiversity and its Survival,* Weidenfeld and Nicholson, London, 1995.

Lovelock, James, *Gaia: A new Look at Life on Earth,* Oxford University Press, 1979.

Lovelock, James, *The Ages of Gaia: A Biography of our Living Earth,* Oxford University Press, 1995.

Lovelock, James, *Healing Gaia: Practical Medicine for the Planet,* Harmony Books, 1991.

Lutzenberger. Jose, A., *Rainforests and World Climate,* Sundance Summit, August 1989.

Moran, Emilio F., *Human Adaptability: An Introduction to Ecological Anthropology,* Westview Press, Boulder, Colorado, 1982.

Nisbet, Euan, *Living Earth: A short History of Life and its Home,* Chapman and Hall, London, 1991.

Parry, Martin, *Climate Change and World Agriculture,* Earthscan, 1990.

Phillips, Oliver *et al,* 'Changes in the Carbon Balance of Tropical Forests: Evidence from long-term Plots,' *Science,* Vol. 282, 16 October 1998.

Pimental, David, *Global Climate Change and Agriculture,* Cornell University, 3 November, 1998.

Retallack, Simon, 'How US Politics is letting the World down,' The *Ecologist,* Vol. 29, No. 2, 1999.

Retallack, S., and Bunyard, P., 'We're damaging our Climate! Who can doubt it?' The *Ecologist,* Vol. 29, No. 2, 1999.

Rosenzweig, Cynthia and Parry, Martin, 'The potential Impact of Climate Change on World Food Supply,' *Nature,* Vol. 367, 13 January, 1994.

Salati, Eneas, 'The Forest and the Hydrological Cycle,' in *The Geophysiology of Amazonia,* Robert Dickinson (Ed), Wiley and Son, 1987.

Schneider, Stephen H., *Laboratory Earth: The planetary Gamble we can't afford to lose,* Weidenfeld and Nicholson, London, 1996.

Schneider, Stephen H., *Global Warming,* Sierra Club, 1989.

Smith, Paul M. and Warr, Kiki (Eds), Global Environmental Issues, Hodder and Stoughton, 1991.

The *Ecologist*, Vol. 29, No.2, March/April 1999, Climate Crisis special issue.

The Hadley Centre, *Modelling Climate Change,* 1995.

The Meteorological Office, *Climate Change and its Impacts,* November 1998.

The Meteorological Office, *Climate Change Scenarios for the United Kingdom,* 1998.

Thompson, Lonnie, Davis, Mary E., and Mosley-Thompson, Ellen, 'Glacial Records of Global Climate: a 1500 year tropical Ice Record of Climate,' *Human Ecology,* Vol.22, No. 1, 1994.

Tudge, Colin, *The Day before Yesterday: Five Million Years of human History,* Pimlico, London, 1996.

Volk, Tyler, *Gaia's Body: Towards a Physiology of Earth,* Springer-Verlag, New York, 1998.

Waugh, David, *Geography: An integrated Approach,* Nelson, 1995.

Zhang, H., McGuffie, K. and Henderson-Sellers, A., *Journal of Climate,* Vol. 9, 1996.

Index